T0192242

Wireless Local Area Networks Quality of Service

An Engineering Perspective

Osama Aboul-Magd

Published by
Standards Information Network
IEEE Press

Trademarks and Disclaimers

IEEE believes the information in this publication is accurate as of its publication date; such information is subject to change without notice. IEEE is not responsible for any inadvertent errors.

Library of Congress Cataloging-in-Publication Data

Aboul-Magd, Osama, 1956–
Wireless local area networks quality of service: an engineering perspective / Osama Aboul-Magd.
p. cm.
Includes bibliographical references and index.
ISBN 978-0-7381-5673-6

1. Wireless LANs. 2. Wireless communication systems. I. Title.

TK5105.78.A23 2007

621.384--dc22

2007061361

IEEE
3 Park Avenue, New York, NY 10016-5997, USA

Jennifer McClain, Managing Editor

Review Policy

The information contained in IEEE Press/Standards Information Network publications is reviewed and evaluated by peer reviewers of relevant IEEE Technical Societies, Standards Committees and/or Working Groups, and/or relevant technical organizations. The authors addressed all of the reviewers' comments to the satisfaction of both the IEEE Standards Information Network and those who served as peer reviewers for this document.

The quality of the presentation of information contained in this publication reflects not only the obvious efforts of the authors, but also the work of these peer reviewers. The IEEE Press acknowledges with appreciation their dedication and contribution of time and effort on behalf of the IEEE.

To order IEEE Press Publications, call 1-800-678-IEEE.

Print: ISBN 978-0-7381-5673 STDSP 1151

See other IEEE standards and standards-related product listings at: http://standards.ieee.org/

Author

Osama Aboul-Magd received the Bachelor of Science degree in Electrical Engineering from Cairo University, and the M.A.Sc and Ph.D from the University of Toronto in 1988. Since then he has been employed at Nortel Networks. At Nortel he has been working on planning and design of QoS and traffic management features for many products, covering multiple technologies including Frame Relay, ATM, IP, and Ethernet. He was one of the main contributors to the ATM Traffic Management specification at the ATM Forum. He was the editor of the Ethernet Traffic Management Specification at the Metro Ethernet Forum. He started attending the IEEE 802.11 WG on May 2003. He is interested in WLAN QoS- and MAC-related issues.

Dedication

To the memory of my parents

Acknowledgments

Behind this book is the work of the larger IEEE 802 community, in particular, the outstanding work performed by the IEEE 802.11 Working Group.

I would like to express my sincere thanks to my colleagues at Nortel Networks, especially Ed Juskevicius, Bilel Jamoussi, and Hesham El-Bakoury for their help and encouragement.

I also would like to thank the anonymous peer reviewers of this book for their evaluation and guidance in helping to enrich the material presented here.

Finally, I would like to thank Jennifer McClain and the rest of the IEEE editorial team for their diligence and the great work performed.

Table of Contents

Preface

Since their introduction, wireless local area networks (WLAN) based on IEEE 802.11 technologies have seen a wide range of applications. A common application is the deployment of WLAN equipment during conventions. Participants log in to their corporate networks to gain access to their email and other data applications, such as file transfer. WLAN has also found its way to the home, where it is now customary to find WLAN access points (AP) to allow convenient access anywhere to the Internet. Enterprises and small businesses are also increasingly deploying WLAN equipment, facilitating unwired connectivity to their employees.

With the introduction of Voice over the Internet Protocol (VoIP) and other multimedia applications, WLAN users demand more than just connectivity based on best-effort delivery. One user requirement is for WLAN to support quality of service (QoS) mechanisms, allowing the differentiation between traffic types and the allocation of resources to application traffic that has to be delivered with performance assurances in terms of delay and loss.

In recognition of the need to support QoS, the IEEE 802.11 working group (WG) started working on WLAN QoS and formed a task group (TG), task group e, to study and develop recommendations on the subject. TGe has concluded its work with the publishing of IEEE 802.11e–2005, which is now incorporated in the new amendment of IEEE 802.11–2007 standards.

This book describes the support of WLAN QoS as specified by IEEE 802.11e. It first introduces the subject of QoS mechanisms and popular QoS architectures that are developed for other technologies such the asynchronous transfer mode (ATM) and the Internet Protocol (IP). It then describes WLAN QoS features and maps those features to architectures introduced before. This mapping is essential to facilitate and explain the interworking between WLAN QoS and that of different technologies, e.g., IP.

The term *QoS* has been used in the industry to mean different things. Some use QoS to denote those network mechanisms such as buffer management and scheduling that are necessary for traffic differentiation. Others use QoS to denote all aspects of the service as observed by the end user. The International Telecommunication Union (ITU-T) introduces in its E.800 recommendation a formal definition for QoS. E.800 defines QoS as the collective effect of service performance which determines the degree of satisfaction of a user of a service. This definition encompasses all aspects of the system, including at the radio level, and is not just limited to QoS mechanisms that are usually referred to as the traffic management mechanisms. This book will use the term *QoS* to refer to traffic management mechanisms, with the understanding that QoS has more dimensions than just traffic management.

QoS support is as old as the subject of networking. The public telephone network was the first to support QoS by assigning a physical circuit (a time slot or a frequency band) for the sole use of a single voice call (a single user). The circuit was reserved and maintained for the entire duration of the call and then released when the call concluded.

The history of QoS in packet networks is more recent. IP and Ethernet, for example, were traditionally managed on a best-effort basis. *Best effort* implies that the network attempts to deliver the information from one point to the another without any commitment that the information will actually be delivered, and if delivered, it will be delivered in a timely fashion. With best-effort delivery, the performance experienced by the user could fluctuate, depending on network conditions.

With the introduction of broadband integrated service network (B-ISDN), based on ATM technology, issues related to packet network QoS have received much attention from industry and academia alike. Progress was made on understanding the different QoS mechanisms, such as shaping, policing, scheduling, etc., and the related deterministic delay bounds. At the same time, the state of the art has advanced greatly as a result of the effort to support QoS-based services on an IP platform.

This book is written with two objectives in mind. The first objective is to explain WLAN QoS standards and provide a rationale for the different design choices, given the characteristics of the wireless media. The second objective is to map WLAN QoS to well-known QoS architectures. This mapping is necessary to facilitate interworking when WLAN is deployed in an environment where different networking technologies are in use with different QoS architectures.

Osama Aboul-Magd

Acronyms

ABR	Available Bit Rate
AC	Access Category
ACK	Acknowledgement
AF	Assured Forwarding
AIFS	Arbitration Inter-Frame Spacing
AP	Access Point
ARQ	Automatic Repeat Request
ATM	Asynchronous Transfer Mode
BA	Behavior Aggregate
BA	Block ACK
BB	Bandwidth Broker
BCM	Backward Congestion Management
BE	Best Effort
BSS	Basic Service Set
CAC	Call Admission Control
CAP	Controlled Access Phase
CBR	Constant Bit Rate
CCA	Clear Channel Assessment
CE	Customer Equipment; also, Congestion Experienced
CF	Contention-Free

CFP	Contention-Free Period
CIR	Committed Information Rate
CLP	Cell Loss Priority
CLS	Controlled Load Service
CRC	Cyclic Redundancy Check
CS	Class Selector; also, Carrier Sensing
CS-CO	Circuit-Switching Connection-Oriented
CSMA/CA	Carrier Sense Multiple Access/Collision Avoidance
CTS	Clear to Send
DA	Destination Address
DCF	Distributed Coordination Function
DE	Discard Eligibility
DF	Default Forwarding
Diffserv	Differentiated Service
DIFS	Distributed Interframe Spacing
DLS	Direct Link Setup
DS	Distribution System
DSCP	Diffserv Code Point
DTIM	Delivery Traffic Indication Map
EBR	Effective Bit Rate
ECN	Explicit Congestion Notification
EDCA	Enhanced Distributed Channel Access
EF	Expedited Forwarding
EIFS	Extended Interframe Spacing
EIR	Excess Information Rate

ER	Explicit Rate
ERP	Extended-Rate PHY
ESS	Extended Service Set
FCM	Forward Congestion Management
FDM	Frequency Division Multiplexing
FIFO	First-In-First-Out
FR	Frame Relay
GPS	Generalized Processor Sharing
GR	Guaranteed Rate
GS	Guaranteed Service
HCCA	HCF Controlled Channel Access
HCF	Hybrid Coordination Function
IBSS	Independent BSS
IE	Information Element
IETF	Internet Engineering Task Force
IntServ	Integrated Service
IP	Internet Protocol
IPTV	IP Television
ISI	Intersymbol Interference
ISM	Industrial, Scientific, and Medical
IWU	Interworking Unit
MEF	Metro Ethernet Forum
MMFP	Markov Modulated Flow Process
MPDU	MAC Protocol Data Unit
MSDU	MAC Protocol Service Unit

MTT	Maximum Theoretical Throughput
MU	Mobile Unit
NAV	Network Allocation Vector
OFDM	Orthogonal FDM
OSPF	Open Shortest Path First
PC	Point Coordinator
PCF	Point Coordination Function
PCP	Priority Code Point
PDF	Probability Density Function
PHB	Per-Hop Behavior
PHY	Physical Layer
PIFS	PCF Inter-Frame Spacing
PLCP	Physical Layer Convergence Procedure
PNNI	Private Network-to-Network Interface
PPDU	PLCP Protocol Data Unit
PS	Power Save
PS-CO	Packet-Switching Connection-Oriented
PSK	Phase Shift Keying
PSTN	Public Switched Telephone Network
PTD	Packet Transfer Delay
QAM	Quadrature Amplitude Modulation
QoS	Quality of Service
RED	Random Early Detection
RR	Round Robin
RSVP	Resource Reservation Protocol

RTS	Request to Send
RTO	Retransmission Time-Out
RTP	Real-Time Transport Protocol
RTT	Round Trip Time
SA	Source Address
SAP	Service Access Point
SIFS	Short Interframe Spacing
SLS	Service Level Specification
SONET	Synchronous Optical Network
SP	Service Period
SS7	Signaling System 7
SSID	Service Set Identifier
STA	Station
TBTT	Target Beacon Transmission Time
TC	Traffic Category
TCP	Transport Control Protocol
TDM	Traffic Division Multiplexing
TID	Traffic Identifier
TIM	Time Indication Map
ToS	Type of Service
TS	Traffic Stream
TSF	Time Synchronization Function
TSID	TS Identifier
TU-T	International Telecommunication Union
TXOP	Transmission Opportunity

UDP	User Datagram Protocol
UP	User Priority
VBR	Variable Bit Rate
VID	VLAN IDentifier
VLAN	Virtual LAN
VoIP	Voice over Internet Protocol
WLAN	Wireless Local Area Networks
WM	Wireless Media
WRR	Weighted Round Robin

Chapter 1 Quality of Service Mechanisms and Metrics

This chapter describes the different traffic management mechanisms commonly employed by communication networks and their impact on network design. A network designer might opt to deploy all or a subset of the mechanisms described here. The choice of mechanisms to be deployed is highly dependent on the performance requirements of the applications being supported by the network. Real-time applications such as voice and video require stringent performance requirements in terms of information loss and delay. Hence every attempt has to be made to ensure that traffic does not exceed levels where application performance can no longer be achieved. On the other hand, a network supporting best-effort service might require a small subset or none of the quality of service (QoS) mechanisms to achieve its mission.

Applications and network performance are measured objectively using performance metrics. Metrics such as delay and loss are commonly used to judge application performance. These metrics and their definitions are also introduced in this chapter.

Traffic Management Mechanisms

Traffic management mechanisms can be classified based on a number of criteria. One possible approach is to classify those mechanisms based on the time scale at which they operate. Figure 1–1 shows a number of traffic management mechanisms relative to their time scales. These domestically range from the time needed to transmit a single packet at the wire speed to the time needed to complete a session or a connection.

Depending on the link speeds and the packet size, the packet time could be on the order of fraction of a microsecond. For packet forwarding to occur at the wire speed, decisions related to buffering and scheduling of incoming and outgoing packets have to be made within the timescale needed to receive or transmit a single packet. Those decisions are usually implemented in hardware.

A traffic management mechanism that requires feedback from network nodes back to the end systems has a timescale on the order of the round-trip propagation delay of the network. Depending on the network geographical coverage, this time could range from few microseconds to a few milliseconds. Mechanisms based on network feedback are usually helpful in alleviating network congestion and are usually referred to as congestion management or congestion control mechanisms.

Traffic management mechanisms such as admission control have an impact that is observable during the lifetime of a session or a connection. This time can range from few seconds, to minutes, to even hours. For example, the average duration of a voice call is estimated to be on the order of 3 minutes.

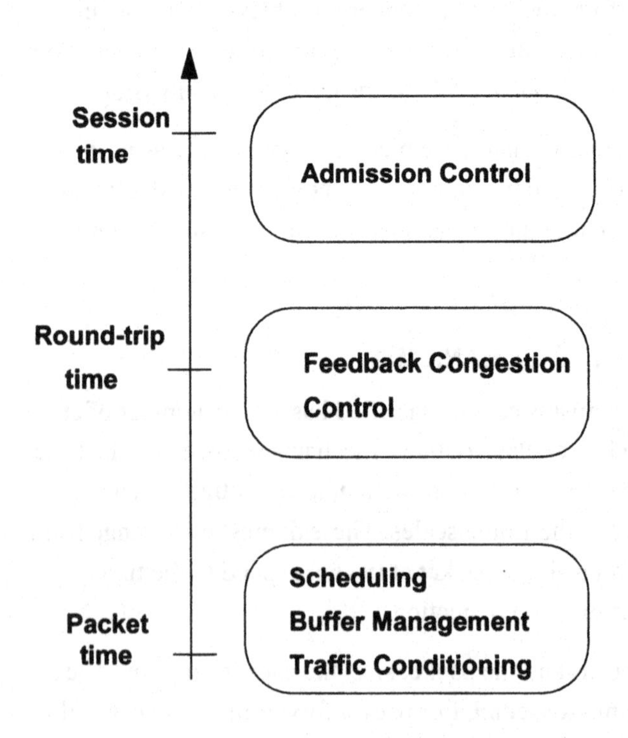

Figure 1–1: Traffic Management Mechanisms and Timescale

In addition to the time scale classification, some of the mechanisms shown in Figure 1–1 are network-wide mechanisms that require cooperation from the different network elements, e.g., admission control. Other mechanisms are nodal mechanisms, such as transmission scheduler and buffer management.

Furthermore, some mechanisms apply only at the edge of the network. Others must operate at every node in the network. For instance, traffic metering and policing (traffic conditioning), as described in "Traffic Conditioning" on page 4, are usually implemented at the edge node. However, transmission scheduling is implemented at every node of the network. Figure 1–2 shows traffic management mechanisms and the locations in the network where they can be implemented. For example, the shaping function can be implemented at points where user equipment interfaces with the network or where the network interfaces with another network.

Figure 1–2 distinguishes between two paths, the data path and the control path. The data path is the path where exchange of data units takes place. The control path is where the exchange of control information, including the signaling required to support the QoS functions, takes place. These two paths

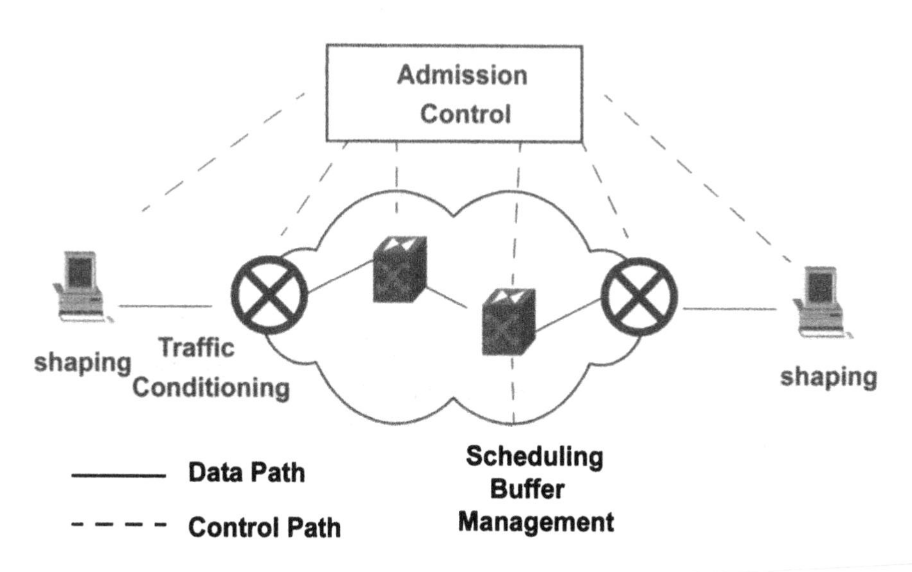

Figure 1–2: Traffic Management Functions and Scope

need not be different, as in Figure 1–2. In many networks, including those based on IP, the data and the control paths coincide. The situation where control information follows the same path as that followed by the data is called *in-path* signaling. Otherwise it is called *off-path* signaling.

TRAFFIC CONDITIONING

Traffic conditioning is a term that collectively describes a group of traffic management functions usually implemented at the network edge. The main function of traffic conditioning is to ensure that user's traffic adheres to a certain traffic pattern that has been mutually agreed upon with the network provider. This traffic pattern is usually referred to as the *arrival curve*. The main functions of the traffic conditioner are shown in Figure 1–3 [B4].

The input to the traffic conditioning function is a stream of information units (packets or frames). The first step in traffic conditioning is to identify subsets of the offered traffic that share common features, e.g., packets that have the same source and destination addresses, using the traffic classification function. The identified subsets are called *flows*. Packets belonging to the same flow are then metered, marked (or remarked), dropped, and/or shaped, based

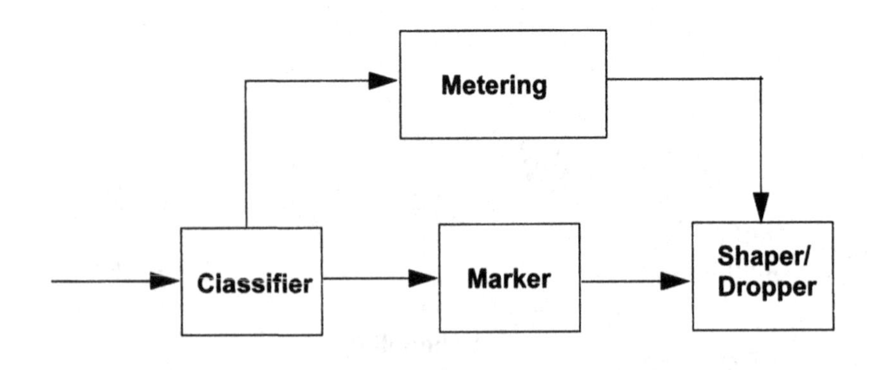

Figure 1–3: Traffic Conditioning Function

on how close a traffic flow adheres to prespecified traffic parameters determined at the admission phase.

The Metering Function

The metering function is sometimes called the *policing* function. The main function of the metering process is to ensure that a traffic flow adheres to some traffic pattern (arrival curve) as defined by some traffic parameters. Those parameters can include a peak information rate and/or a sustained information rate defined over a specified time interval. A meter will characterize each packet as either conforming or nonconforming relative to the traffic parameters characterizing the flow. A conforming packet is determined to be within the flow-specified pattern. A nonconforming packet violates the traffic pattern of the flow as determined by the metering algorithm.

Traffic parameters of a particular flow are subject to negotiation between the end user and the service provider and are included in the Service Level Specification (SLS). A service is usually characterized by a small set of traffic parameters that usually include an information rate and maximum burst size that allows the customer to exceed its defined rate for a well-defined period.

Leaky Bucket

A popular algorithm for implementing the metering function is the one based on leaky bucket implementation. A schematic of a leaky bucket is shown in Figure 1–4. The bucket fills up to a maximum value as determined by the bucket size b. The fill rate depends on the length of the incoming information units and the arrival rate of the incoming flow.

The leaky bucket leaks at a constant rate of r units/s. Incoming packets are said to be (r, b) compliant only if there is a space in the bucket to accommodate the whole packet upon its arrival instant. Figure 1–5 shows a flow chart describing the operation of the leaky bucket. Figure 1–5 shows a bucket implementation where leakage of the bucket is triggered by packet arrival event. Other implementations are also possible.

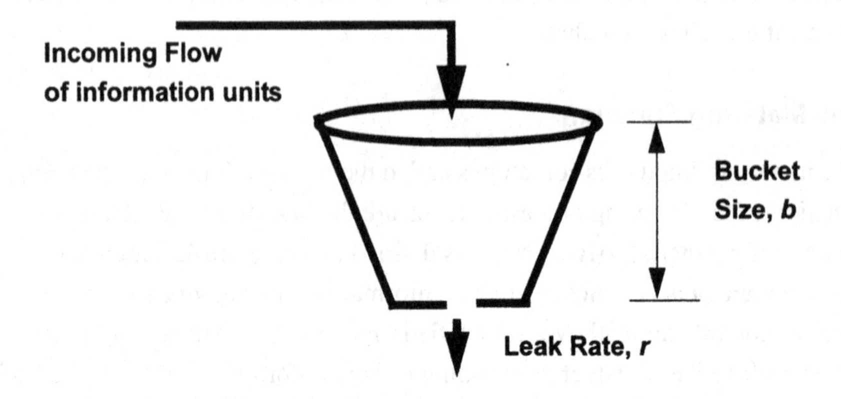

Figure 1–4: Leaky Bucket Schematic

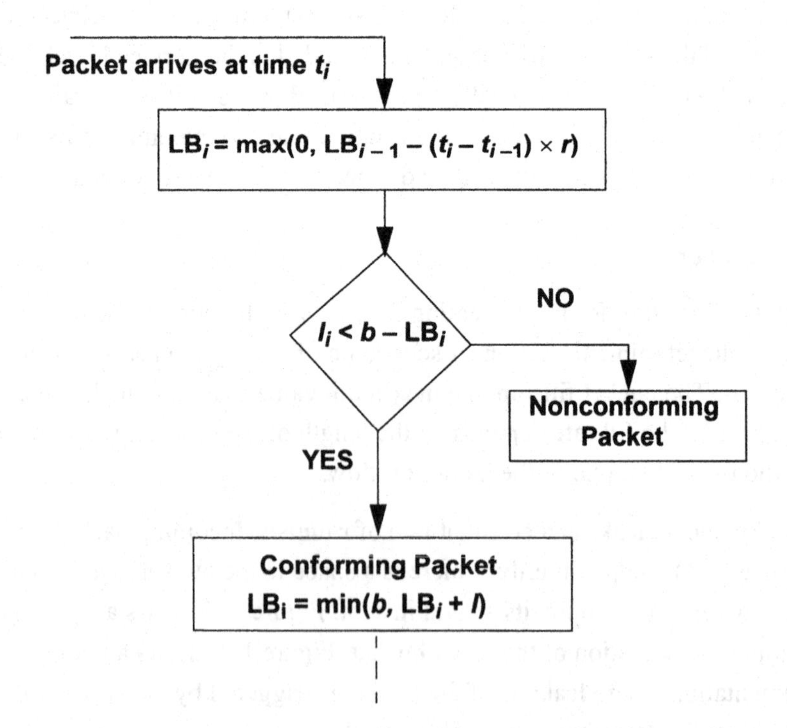

Figure 1–5: Leaky Bucket Flow Chart

An arriving packet of length l_i units (bytes or bits) at time t_i will cause the leaky bucket to leak at the specified rate, r. Then a test is conducted to ensure that the bucket has enough space to admit the incoming packet. If there is enough space, the incoming packet is declared conformant, and the length of the packet will be added to the contents of the bucket LB_i at time t_i. Otherwise, the packet is declared nonconformant.

A single leaky bucket is sufficient to examine traffic conformance to a single rate and a single burst size. More than one leaky bucket arrangement can be put together in order to examine a flow conformance to more than one rate.

A popular arrangement that is in wide use is shown in Figure 1–6, where incoming packets are tested for conformance against a committed information rate (CIR) and an excess information rate (EIR). This arrangement has been used for many years in the context of the Frame Relay (FR) service. It is described in more detail in CCITT I.370 [B6]. It is also the basic algorithm used to define Metro Ethernet Forum (MEF) services [B25] and for provider backbone bridges as described in IEEE Standard 802.1ad [B14].

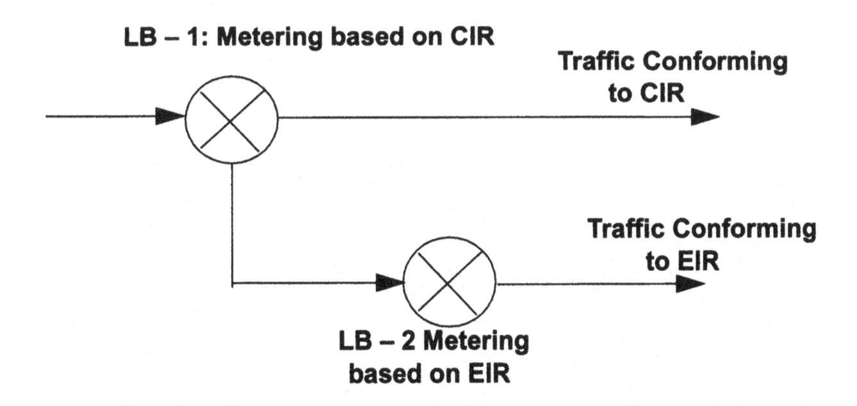

Figure 1–6: Multiple-Leaky Bucket Arrangement for Enforcing Multiple Traffic Rates

A leaky bucket algorithm imposes an upper bound (or an envelope) on the amount of traffic admitted to the network at any interval of length T. Ignoring the finite size of incoming packets and applying a fluid flow approximation, the volume of the conformant traffic X in bits submitted to the network at any time interval of length T is bounded by [B24]:

$$X \leq b + rT \quad \text{EQ 1–1}$$

Equation 1–1 defines the arrival curve specified by the leaky bucket with parameters r and b. Figure 1–7 shows the behavior of X as a function of time. Knowledge of an upper bound is helpful in allocating network resources to satisfy application performance requirements. Bounds on the offered traffic make it possible to estimate an upper bound on the worst-case delay observed by a particular flow given its traffic envelope.

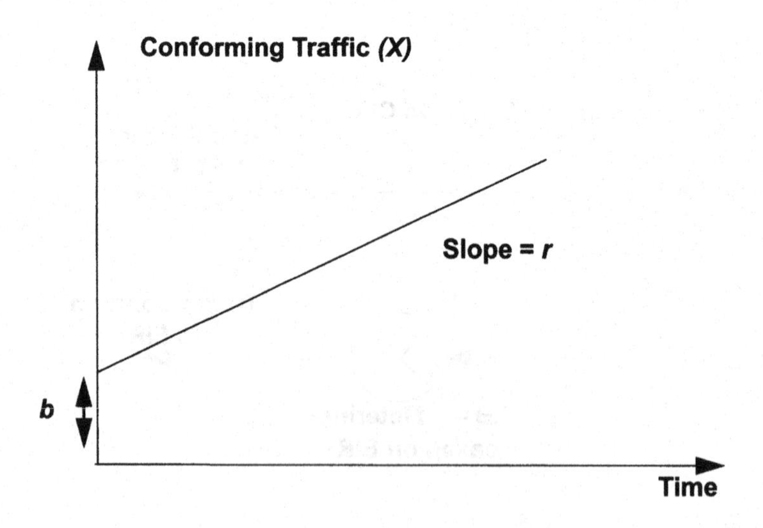

Figure 1–7: Leaky Bucket Limit on Conforming Traffic

Marker

The marking function is concerned with the setting of the appropriate field in the packet header to indicate the kind of treatment the network extends to this packet in terms of buffering and scheduling. Based on classification decisions, a flow of packets will be assigned to a particular transmission scheduler based on the marking in its header. Different application traffic will be marked differently. For instance, real-time flows such as voice and video will be marked to indicate a treatment by the transmission scheduler at different network nodes in such a way to satisfy their stringent delay requirements. On the other hand, non-real-time flows can be delayed without impact on the application performance.

Some aspects of the marking function are tightly coupled to the metering function. Packets that are declared nonconforming by the metering function can be remarked for discard precedence or discard eligibility (DE). Discard eligibility of a particular packet is indicated in a special field of the packet header. IP Differentiated services, Frame Relay DE bit, and ATM cell loss priority (CLP) bit are all examples of networking technologies that support drop precedence based on marking a special field of the packet header [B2]. Packets marked for discard eligibility are discarded upon arrival when the network is close to being congested. The discard of the DE packets can be set at a lower buffer threshold than the one used for discarding other non-DE packets. Discard eligible packets are commonly referred to as *yellow packets*. Non-DE packets are commonly referred to as *green packets*.

Dropper

The dropper function of the traffic conditioning is invoked when there is the need to drop any of the incoming packets. Incoming packets might be dropped for a variety of reasons. For traffic management purposes, a packet might be dropped when it is not conforming to any of the traffic parameters of the flow it belongs to. Dropping, rather than admitting nonconforming packets, is essential to facilitate the engineering of the network resources to satisfy application performance. While nonconforming packets might be allowed to the network and transmitted without assurances, they still consume scarce network resources, and they add uncertainty to the network engineering because there is no prespecified limit on their traffic volume. Packets that are marked for dropping are referred to as *red packets*.

Shaper

The shaping function receives an input traffic stream and forces it to adhere to a prespecified traffic pattern at its output. The output pattern is determined by a shaping curve [B24]. One possible shaping curve is shown in Figure 1–7. This curve forces the incoming traffic to obey leaky bucket constraints at the shaper output.

To satisfy the pattern defined by the shaping curve, each incoming packet will have a play-out time. The play-out time is defined as the time when it is appropriate to send the packet at the shaper output without violating the pattern imposed on the traffic by the shaping curve. Accordingly, a shaper employs a buffer to store incoming packets, if necessary, until their play-out times.

A shaper is inherently a work-nonconserving network element in the sense that the server can stay idle while packets are stored in the buffer [B22]. Figure 1–8 shows a sketch of a shaper that produces an output traffic stream at a constant bit rate of R units/s (assuming all packets are of equal length of L units). Let a_i and d_i represent the arrival and the departure times of the ith packet. The ith packet is stored in the buffer for a time that is equal to:

$$\Delta_i = max\left[0, d_{i-1} + \frac{L}{R} - a_i\right] \quad \textbf{EQ 1–2}$$

The shaping function can be implemented anywhere in the network. In an attempt to reconstruct and recover the initial traffic characteristics in terms of burst size and rate, a network administrator might implement the shaping function when the traffic of some flows traverse certain number of nodes. However, the shaping function is most likely to be found at the interface

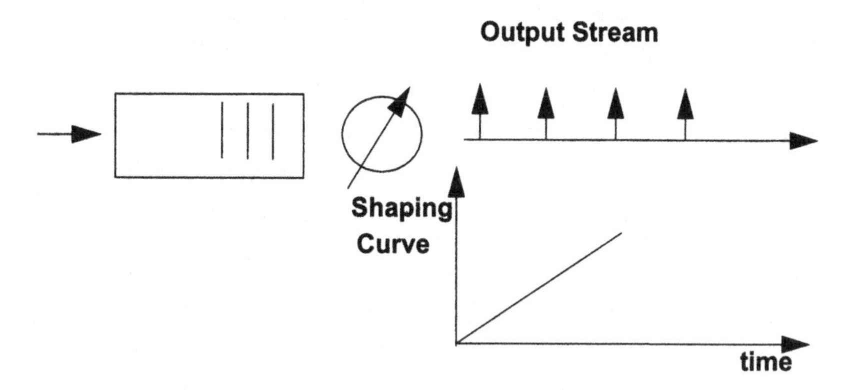

Figure 1–8: Constant Rate Shaper

between the user and the network (user-network interface, UNI) or between two networks (network-to-network interface, NNI). The role of the shaping function in this case is to regulate its output traffic to be within the traffic envelope imposed by the metering or policing function at the other side of the interface. Therefore, the shaping curve in this case coincides with the limits imposed by the metering algorithm.

Admission Control

The main purpose of admission control is to limit the number of sessions admitted to the network and ensure that newly admitted sessions will not negatively impact the performance of the already existing sessions.

Call Admission Control (CAC) has been the main traffic management mechanism employed for PSTN (Public Switched Telephone Networks). Call-blocking probability was the main engineering parameter. Network engineering was mainly related to evaluating the number of circuits needed to meet customers' demand in a particular area. Call-blocking probability was estimated using the Erlang-B formula introduced in Equation 1–38 on page 41.

PSTN CAC was relatively simple to implement and characterize mathematically because all telephone calls (or sessions) have the same characteristics and occupy the same transmission resources, i.e., a homogeneous environment. In PSTN the admission control function on a trunk reduces to making sure that the number of voice calls admitted does not exceed the number of circuits available on the line. Those circuits can be implemented using time division multiplexing (TDM) as in T1 systems and SONET (Synchronous Optical Networks) or frequency division multiplexing (FDM).

With the present networking environment (e.g., IP or WLAN networks), the homogeneous nature of the PSTN is not preserved, and sessions with different resource requirements and different traffic patterns are expected to share the same network resources. The admission control function in a heterogeneous environment is more challenging. A new request should only be admitted if there are sufficient resources to accommodate its needs. That is, the admission of the new request will not negatively affect the performance of the currently ongoing sessions.

To achieve this goal and keep the complexity of the admission control algorithm manageable, the admission control function makes use of certain traffic parameters for admission. Those parameters usually include peak rate and the average rate. Peak rate γ is defined in such a way that over certain period of time of duration T, the amount of information generated during this time is always less or equal to γT. The average rate is the amount of information generated during a period of time divided by the its duration. The time period might extend to the whole duration of a session. The time period over which the average rate is measured is set at a longer value than that used to measure the peak rate.

The straightforward way for admission is to admit new sessions based on their peak bandwidth requirements. However, because most applications generate bursty traffic with an average rate that might be substantially lower than the peak rate, admission strategy based on peak rate allocation will result in a poorly utilized network and hence a waste of network resources.

An alternative way is to admit new sessions based on the long-term average rates of the sessions. Again this admission policy is simple to implement, provided that knowledge of the session average rate is available beforehand. However, average allocation could result in performance degradation during periods where traffic rate deviates substantially from its average value.

A compromise between these two strategies is to define an effective bit rate (EBR) approach where session requirements in terms of transmission resources are determined as a function of the session traffic parameters (peak and average rates) and its performance requirements (packet loss and delay objectives). The EBR is always bounded as:

$$\omega \leq EBR \leq \gamma \qquad \textbf{EQ 1–3}$$

where γ is the session peak rate and ω is its average rate.

A new request will be admitted when the sum of all the EBR of all sessions in progress and the new request is less than the link-available bandwidth. An example of EBR calculation that has seen some applications, especially in the context of ATM networks, is that proposed by Gibbens and Hunt [B11] for two-state Markov Modulated Flow Process (MMFP). The two-state MMFP traffic model is shown in Figure 1–9.

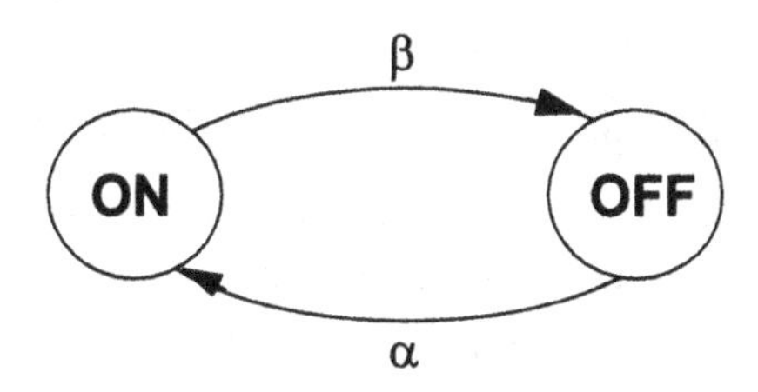

Figure 1–9: MMFP Transition Diagram

The MMFP traffic source alternates between ON and OFF periods. During ON periods information is generated with rate γ units/s. No information is generated during the OFF periods. The durations of the ON and OFF periods are assumed to have exponential distributions with parameters β and α, respectively. Accordingly, MMFP traffic sources generate traffic at an average rate that is equal to:

$$\omega = \gamma \times \frac{\alpha}{\alpha + \beta} \qquad \textbf{EQ 1–4}$$

With these assumptions, the EBR calculated using Gibbens-Hunt [B11] procedure is given by:

$$EBR = \frac{-[\alpha + \beta + \theta\gamma] + \sqrt{[\alpha + \beta + \theta\gamma] + 4\theta\alpha\gamma}}{2\theta} \qquad \textbf{EQ 1–5}$$

where θ is the packet loss objective normalized by the available buffer size B:

$$\theta = \frac{Ploss}{B} \qquad \textbf{EQ 1–6}$$

Figure 1–10 shows the behavior of the number of admitted sessions as a function of the ratio of the session peak rate γ and the link rate R (γ/R). The link rate is the capacity of the transmission facility over which sessions are multiplexed. The number of admitted sessions is computed by dividing the link rate by the rate allocated for each session.

When γ/R is small, the number of admitted sessions is close to the number of admitted sessions based on Average Rate allocation. As γ/R increases, the number of admitted sessions deviates from the Average Rate allocation line and approaches the one based on Peak Rate allocation.

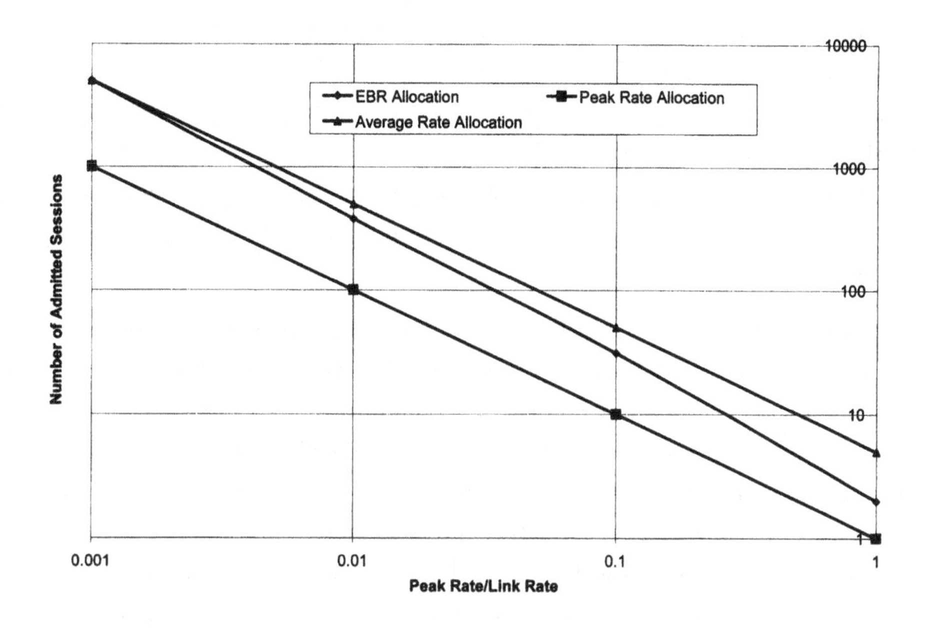

Figure 1–10: EBR Behavior

Nodal Mechanisms

Nodal resources in terms of buffer space and transmission capacity are usually shared among a number of flows. An assured QoS offering necessitates the dedication and sharing of resources between individual flows based on their QoS objectives. The dedication and sharing of resources are achieved by employing a transmission scheduler and buffer allocation policies.

Transmission Scheduler

The transmission scheduler's function is to select packets belonging to different flows for transmission at different times, with the objective of ensuring that the rates assigned to different flows are achieved within a certain error bound. The error bound is a function of the scheduler design.

The study of rate schedulers and their related performances was an active area of research during the time when the notion of the guaranteed QoS was discussed in the context of ATM traffic management [B2] and IP Integrated Services (IntServ) [B31]. A rigorous mathematical treatment of the subject is available in reference books [B24] and a large number of research papers. The discussion here attempts to simplify the mathematical argument without sacrificing accuracy. For more rigorous presentation, refer to one of the references mentioned.

A guaranteed rate scheduler attempts to isolate the different flows and assign for each flow the rate allocated to it. In a perfect world the isolation of flows is achieved by allocating a separate queue for each flow. Each queue is then served by a transmission rate equal to the rate assigned to the corresponding flow, as shown in Figure 1–11. Practical implementation of the arrangement in Figure 1–11 requires the partition of the transmission resources to portions with different rates. This partition might be possible in circuit-switched networks but not feasible in the packet-switched networks that prevail in the current network environment. The statistical multiplexing nature of packet networks inherently allows resource sharing. Statistical multiplexing makes packet-switched networks more efficient than circuit-switched networks, where sharing is not allowed and hard to implement.

The arrangement shown in Figure 1–11 is a possible representation of the generalized processor sharing (GPS) introduced by Parekh and Gallager [B28]. GPS is an idealized scheduler. It assumes the server is acting on all backlogged flows simultaneously in such a way that the portion of the service capacity assigned to the ith flow in an interval of length τ is given by:

$$S_i(t, t+\tau) \geq \frac{\phi_i}{\sum_j \phi_j} \qquad \textbf{EQ 1–7}$$

where ϕ_i is the fraction of the transmission bandwidth assigned to the ith flow.

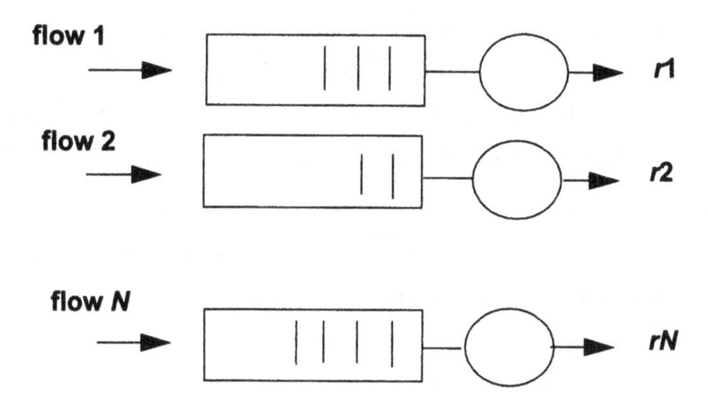

Figure 1–11: Flow Isolation

With the ith flow in isolation, its assigned rate is denoted by $r_i = \frac{\phi_i}{\Sigma\phi_j}C$, for $i = 1,2,...,N$, where C is the transmission capacity in bits per second, and N is the number of admitted flows. Let a_k, d_k, and L_k in turn stand for the arrival time (defined as the time the last bit of the packet arrives at the queue), departure time (defined as the time instant at which the last bit of the packet leaves the queue), and the length of the kth packet of a given flow. A packet arriving at an empty system (no packets exist in the queue or are being served) will proceed directly to the service process and will depart after the time needed to transmit the whole packet. This time is given by $\frac{L_k}{r_i}$. On the other hand, a packet arriving at a nonempty queue has to wait until the departure time of the previous packet, d_{k-1}, before advancing to the service process. It leaves the queue $\frac{L_k}{r_i}$ s later. To capture these two distinct cases, the relationship between the departure times of the different packets of the ith flow can be expressed as:

$$d_k = max(a_k, d_{k-1}) + \frac{L_k}{r_i} \qquad \text{EQ 1–8}$$

This relationship is derived assuming the ideal case that the transmission resources can be partitioned between the different flows entering the node, as shown in Figure 1–11. This assumption is equivalent to assuming an idealized GPS scheduler.

In a realistic scenario, this clear partition is not feasible, and packets from different flows share the same transmission facility. To account for such sharing and to accommodate cases where a packet has to wait out the time needed to finish the transmission of another packet that might belong to a different flow, we introduce the following definition [B12].

Definition: A transmission scheduler is said to be a *guaranteed rate* (GR) scheduler if it guarantees that a packet will be transmitted within ε of its departure time as given by the expression for d_k. The ε term is an error term that depends on the scheduler design.

The kth packet of the flow under consideration traversing a series of nodes on its way from source to destination will encounter an end-to-end delay, D_k, that is bounded by:

$$D_k \leq d_k - a_k + (M-1) \times max\frac{L_n}{r_i} + \sum_{j=1}^{M} \alpha^j \qquad \text{EQ 1–9}$$

where α^j is equal to the sum of the error terms ε^j and the propagation delay, τ^j between the j and the $j + 1$ node.

M represents the number of nodes traversed by the packet on the way to its destination. The delay inequality has three terms. The first term is the normal delay of a packet if it arrives at a single queue at a_k and departs at d_k. The second term assumes a worst case where the packet under consideration is always of the maximum possible length at every node along the path. The third term depends on the sum of the error terms due to scheduler architectures at the different nodes along the path.

For a flow that is regulated by a leaky bucket with parameters (r, b), this delay bound reduces to:

$$D \leq \frac{b + (M-1) \times L_{max}}{r_i} + \sum_{j=1}^{n} \alpha^j \qquad \text{EQ 1–10}$$

This relationship is necessary to ensure a deterministic upper bound on the delay performance of flows and is useful in characterizing guaranteed services with delay assurances. The next section describes a number of popular transmission schedulers and characterizes their performance as given by the Equation 1–8, Equation 1–9, and Equation 1–10.

Transmission Scheduler Examples

In this section different types of transmission scheduler are presented. The objective here is not to present all scheduler types that have been discussed in the literature but rather to select those types that, to the best of our knowledge, have seen wide implementation in actual products.

First Come First Serve (FCFS) Scheduler

In a FCFS service discipline, all incoming flows are aggregated in a single flow in which packets are served in the order of arrival, as shown in Figure 1–12. The FCFS discipline is not a guaranteed rate scheduler in the sense that it has no ability to guarantee a rate for a particular flow. Furthermore, it has no ability to discriminate against incoming packets to give a favorable treatment for those packets that belong to real-time flows such as voice and video.

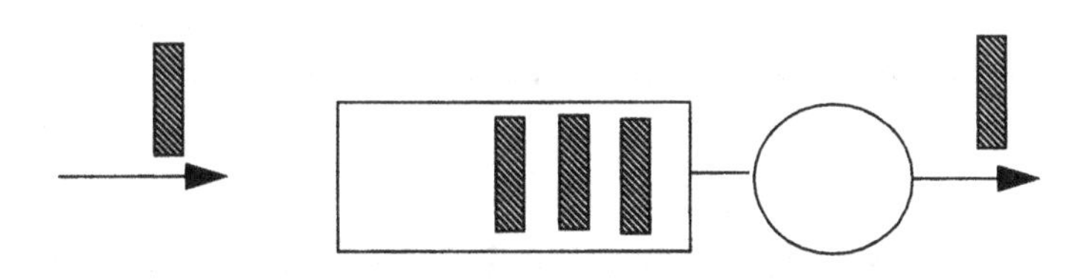

Figure 1–12: FCFS Service Discipline

Absolute Priority Scheduler

In an absolute priority scheduler, flows are aggregated into one of P aggregate flows according to the priority assigned to them. Queues are then served in a priority fashion so that the ith queue is served before the $(i-1)^{st}$ queue. Figure 1–13 shows the case where $P = 2$, i.e., high- and low-priority classes. The service process of the low-priority queue will start only when the high-priority queue is empty. Figure 1–13 shows the output packet sequence for the queue occupancy depicted. The two high-priority packets will leave first before the server starts transmitting low-priority packets.

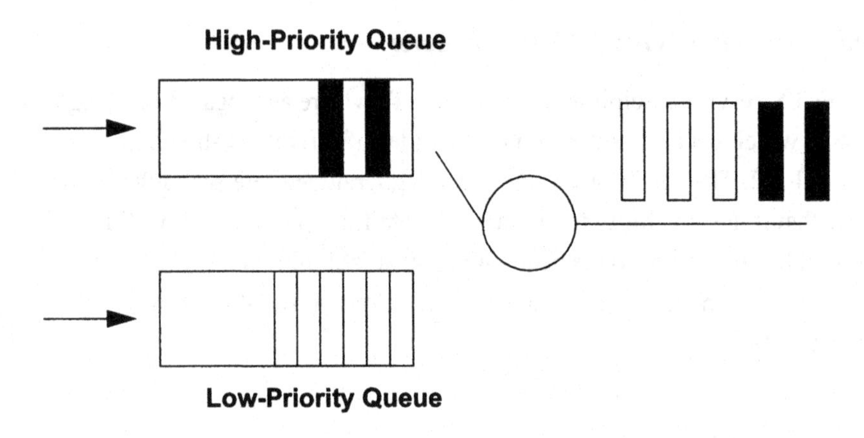

Figure 1–13: Absolute Priority Schedule

Like the FCFS discipline, the absolute priority scheduler is not a guaranteed rate scheduler, particularly for the cases where the high-priority traffic is left unregulated. However, unlike the FCFS discipline, the absolute priority scheduler can discriminate between flows and, when kept at appropriate loading, it will offer superior delay performance for real-time flows. Equivalently, the highest priority class will have a guaranteed rate equal to the transmission capacity with an error term, ε, that is equal to L_{max}/C.

Round Robin (RR) Scheduler

An RR scheduler employs N queues, one for each of the flows it supports. In this service discipline, a packet is transmitted from the head of a queue before the server moves to transmit a packet from the next queue. A service cycle will be completed when a packet is served from each of the nonempty queues, and the service process returns back to its starting position. Figure 1–14 shows the basic idea for an RR scheduler serving two queues ($N = 2$) and the corresponding output sequence for the queue occupancy depicted.

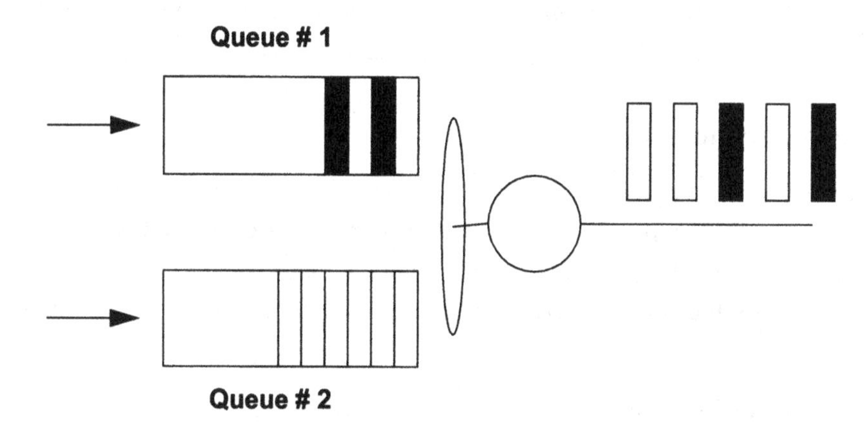

Figure 1–14: Weighted Round Robin (WRR) Schedule

An RR service scheduler is a guaranteed rate scheduler with $\varepsilon = \sum_{i=1}^{N} \frac{L_{max}}{C}$, where L_{max} is the maximum size packet allowed. It is worth noting that the parameter ε is calculated assuming worst case, where every queue always has a packet of maximum length waiting for transmission. When the number of flows N increases, ε becomes so increasingly large that the delay bound becomes prohibitively large and almost meaningless.

Because packets are generally of variable size, an RR scheduler cannot ensure fairness between the different flows. A flow with long packets will be able to capture a bigger share of the transmission bandwidth than a flow with short packets. Fairness would be achieved in cases where all packets for all flows are the same length.

A nontrivial variation of the RR scheduler is the weighted RR (WRR) scheduler. WRR scheduler allows for a finer differentiation between different flows by assigning different portions of the transmission resources to different flows. At each visit the WRR scheduler will serve a number of packets from a queue different from the number served from other queues, depending on the bandwidth allocation for flows sharing the transmission facility.

Buffer Management

The main two issues related to nodal buffer management are the buffer sharing and packet drop policies. A buffer-sharing policy is concerned with the sharing of the available storage space among the many sessions (or flows) competing for the nodal resources. A number of sharing policies are possible, including the following arrangements:

- **Complete Partition**: This is a buffer management policy where the buffer space available is allocated equally among the flows sharing the nodal resources. An active flow is not allowed to make use of unused buffer space allocated for other sessions. With this strategy, if the buffer size is B_{max} and N is the number of flows, then each flow will be assigned B_{max}/N buffer space for its exclusive use.
- **Complete Sharing**: This is a buffer management policy where each flow is allowed to occupy a buffer space up to the maximum available, B_{max}. Complete sharing policy allows a better buffer utilization, but it does not guard against sessions that might flood the buffer, either intentionally or unintentionally. In these cases, denial of service for well-behaving sessions is a possibility (denial of service attack).
- **Sharing with Maximum Allocation**: To guard against the denial of service attack that may occur with the complete sharing policy, an allocation

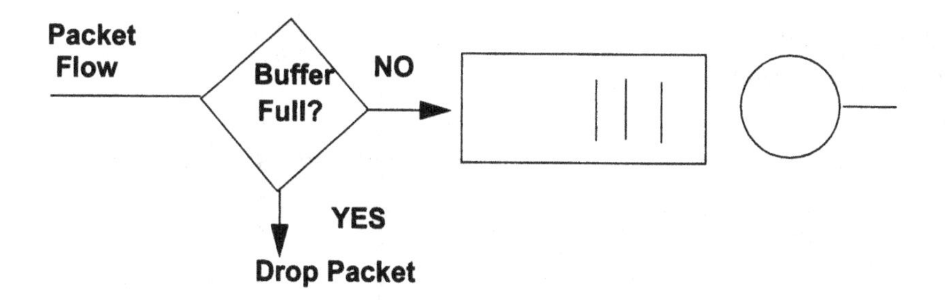

Figure 1–15: Tail-Drop Policy

policy could be used where each session is allocated a maximum limit B_i so that $B_i < M_{max}$ and $\Sigma B_i > B_{max}$. This policy would ensure that a single session would not be able to monopolize the entire buffer space in its favor, starving other sessions. A square-law allocation method is perceived to be optimal (or close to optimal) where $B_i = \sqrt{B_{max}}$ for all i [B10].

Packet drop policies have impact on the type of service offered to the end system. A simple drop strategy would drop packets on arrival when the buffer is full and there is no space to accommodate the incoming packet, as shown in Figure 1–15. This policy is usually referred to as the *tail-drop* policy.

The tail-drop policy is not the only packet drop policy available, but it is the easiest to implement. Other packet drop policies include *front-drop,* where the packet at the head of the queue will be dropped when the buffer is full. It is also possible to drop a packet anywhere in the queue when the buffer is full, but this policy may be hard to implement.

You can also associate a time-out value for each packet. Then the packet is dropped when the timer expires. Timer-based drop policies are widely implemented for Ethernet-bridged networks based on IEEE 802.1Q work [B15].

A drop policy similar to the tail-drop might employ a number of thresholds in such a way that packets with different discard precedence levels are dropped as the buffer size gradually exceeds these thresholds. Figure 1–16 shows two discard thresholds supporting two discard precedence levels. Packets can then be classified as discard-eligible or normal, based on some criterion. The

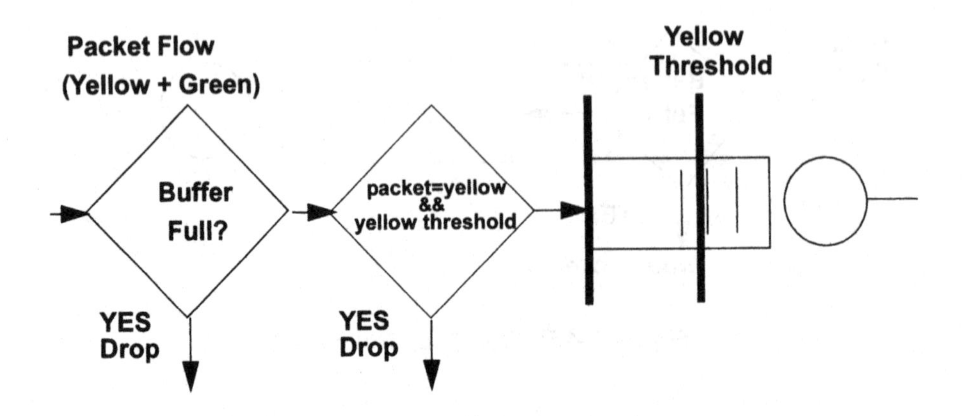

Figure 1–16: Discard Precedence Levels

criterion that is often used is the conformance to traffic parameters as determined by the metering algorithm.

Packets marked for discard eligibility (yellow packets) are discarded before the discard of the non-DE (green) packets when the first threshold is crossed. This mechanism for dropping yellow packets implicitly assumes that the queue size is used as an indication of the nodal congestion status. Discard-eligible packets are dropped when the nodal queue is in danger of being congested. Green packets are only dropped when there is no sufficient buffer space to hold the entire packet.

Tail-drop and threshold-drop policies are ways of passively managing the buffer. The only event that could trigger a packet loss is the exhaustion of the buffer space allocated for certain type of traffic. An active queue management policy will start taking drop actions before the buffer occupancy reaches its maximum.

Active queue management mechanisms such as random early detection (RED) are useful for flows employing TCP as their transport protocol [B9]. RED and its many variations were proposed as a mechanism for throughput improvement when multiple TCP sessions are sharing network resources. The improvement in the performance is achieved by introducing random packet drops to different flows. The random drop has the effect of overcoming flows synchronization commonly observed when a simple tail-drop is employed.

Flows synchronization refers to the phenomenon where the congestion windows of the different TCP sessions decrease and increase simultaneously, causing the network to oscillate between periods of low utilization and periods of overload, where information loss occurs and further reduction in TCP throughput is inevitable.

RED is commonly implemented by computing a running average of the queue length at a particular node and interface. Two thresholds are defined for the queue fill, max_{th} and min_{th}. No action is suggested as long as the average queue fill remains below min_{th}. When the average queue fill is between min_{th} and max_{th}, the incoming packet will be dropped with probability p. The drop probability increases linearly with the average queue fill, as shown in Figure 1–17, until it reaches its maximum value, max_p. When the average queue length exceeds max_{th}, an arriving packet will always be dropped.

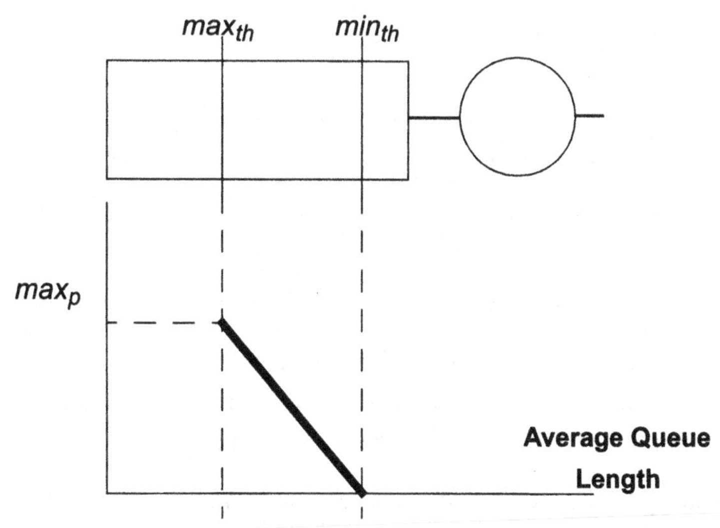

Figure 1–17: Random Early Detection (RED) Model

Feedback Congestion Management

Feedback congestion management requires network nodes to inform end systems of their congestion status. End systems will then adjust their traffic levels accordingly. The feedback may be explicit or implicit.

In explicit feedback, the network may use a special control message or a header field in data packets to convey congestion information back to the end system. Explicit congestion notification (ECN) proposed for IP networks [B29] uses two bits in the IP packet header to indicate to the end system whether congestion was experienced. On the other hand, with implicit congestion notification, the end system recognizes the congestion status of the network by observing congestion symptoms such as long delays and excessive losses. A prominent example of an implicit congestion notification scheme is the TCP congestion-avoidance mechanism.

Figure 1–18 shows the components of a feedback congestion control scheme. The main components are:

- A traffic source where traffic originates and packets are generated. A traffic source can be, for example, a desktop computer supporting VoIP applications or performing file transfer with FTP.
- A reaction point where the information rate is adjusted according to the network congestion status.
- A congestion point where some network resources are shared and could become congested when traffic demand exceeds the available capacity.
- A reflection point where congestion messages are reflected back to the reaction point for an action.
- A traffic destination where source traffic is directed and consumed.

These entities are logical and need not be separated, as shown in Figure 1–18. For example, it is often the case that the traffic source is the place where rate adjustment takes place and hence becomes the reaction point too.

Depending on the location of the reflection point, two classes of feedback congestion control can be identified—backward congestion management

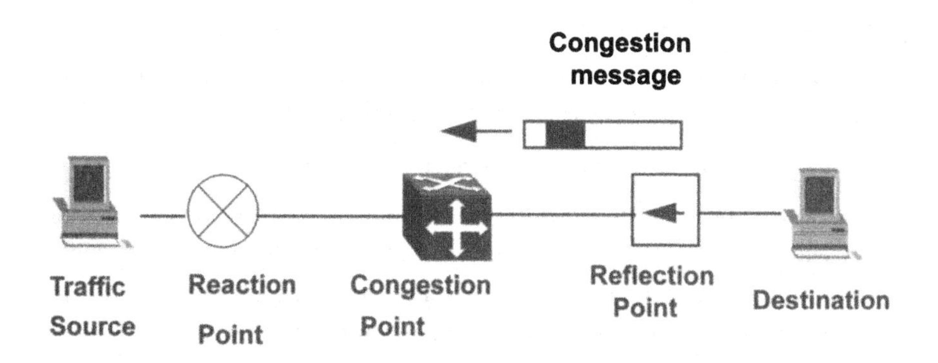

Figure 1–18: Feedback Congestion Control Components

(BCM) and forward congestion management (FCM). With BCM the reflection point coincides with the congestion point. Therefore, when congestion is first detected, it is communicated back to the reaction point. It has the advantage of shortening the control loop delay. However, it does not reflect the congestion status of the whole path. It is safer to use BCM to indicate the need for a traffic decrease to the reflection point. Using BCM for traffic increase can cause problems with downstream congestion points (switches or router).

With FCM the reflection point usually resides with the destination. Hence the congestion message is reflected after it has travelled the whole path between the source and the destination. Therefore, the reflection point has a better view of the congestion status along the whole path before it reflects it back to the reaction point for control action. The delay of the feedback loop in this case is always on the order of a single round-trip delay (RTT)—longer than the feedback delay encountered using BCM. However, FCM is safer to use than BCM because FCM conveys a snapshot of the whole path.

A feedback congestion management scheme that is widely deployed is the one designed for the transport control protocol (TCP) commonly used for IP networks. TCP congestion avoidance mechanism [B18] employs an adaptive window flow control where the window size adjustment is performed at the traffic source (the host). As with any automatic repeat request (ARQ) mechanism, the TCP window size at any time defines the number of TCP bytes that can be transmitted without receiving an acknowledgment (Ack). The current window size is referred to as the congestion window (cwnd). The window size is measured in bytes because TCP is a byte-oriented protocol.

At the onset of the packet exchange, the TCP window behavior follows a slow start phase with cwnd initially set to one TCP segment. Every successful transmission doubles the size of the current window size (exponential growth) effectively doubling cwnd with every round-trip time (RTT). The doubling of cwnd will continue until it reaches a threshold called the slow start threshold (ss_threshold). At this stage the TCP window enters the congestion avoidance phase where the window size is incremented by one segment for every successful transmission, i.e., every RTT (linear growth). When the window size reaches its maximum value, no further increments are allowed. The maximum window size is determined by the receiver-advertised window (rwnd) and is agreed to during the session setup.

The size of the TCP window decreases when congestion is detected by the end system. The end system uses the absence of Ack packets and timeout to detect network congestion. There is no explicit congestion management sent by the network to the host. Upon detecting congestion the end system reduces its congestion window (cwnd). Several variants of TCP differ in their reaction to congestion. TCP Tahoe collapses the cwnd down to one segment, and TCP Reno reduces the size of the congestion window by one-half its current size. The value of the ss_threshold will also be reduced to half its current value.

The TCP source will then enter an exponential backoff phase where retransmission time-out (RTO) doubles for every packet loss until the successful reception of an Ack packet acknowledging the reception of the expected sequence number. The TCP adaptive window will then enter a slow start phase followed by the congestion avoidance phase as discussed before. The behavior of the TCP window is shown in Figure 1–19.

The TCP congestion avoidance mechanism is an example of mechanisms where the reflection point shown in Figure 1–18 either does not exist or its effect is implicitly recognized by the traffic source due to the lack of Ack

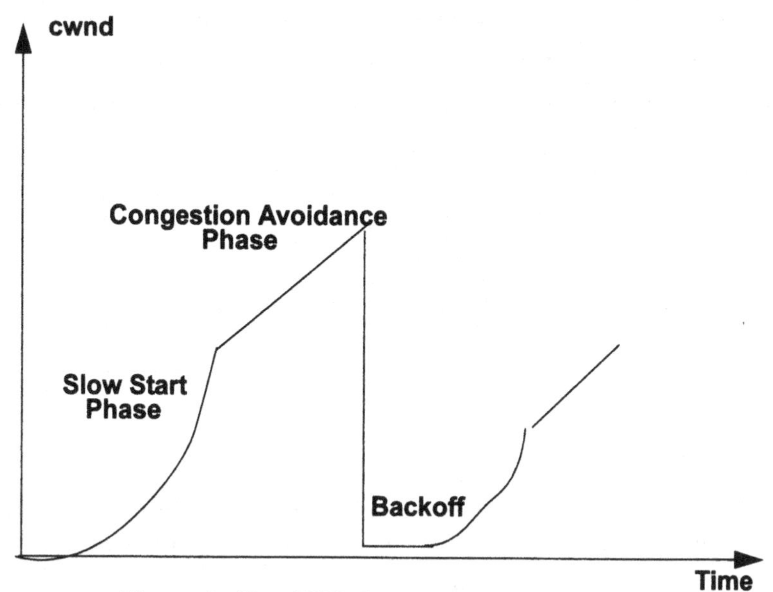

Figure 1–19: TCP Congestion Window

packets. The congestion point in the TCP model is any network element (router) that is susceptible to congestion and packet drop.

The throughput of the TCP congestion avoidance mechanism depends on the network loss probability and the round-trip time (RTT). The relationship between the throughput and these parameters is given by:

$$\gamma = \frac{aMTU}{RTT\sqrt{p}} \qquad \textbf{EQ 1–11}$$

where MTU is the average size of the TCP packet, p is the loss probability, and the parameter a is a constant that can be set at 1.22.

This relationship was obtained assuming stationary network conditions and is accurate for small values of the loss probability in the range of 5%. As p increases, Equation 1–11 tends to overestimate the throughput achieved by TCP and can be interpreted as an upper bound on the actual throughput.

Padhye et al. [B27] offer a more accurate expression for the TCP throughput that takes into account the impact of the maximum window size (W_{max}) and the retransmission time-out (RTO). They developed the following equation.

When the average window size $E[W]$ is less than W_{max}:

$$\gamma = \frac{\frac{q}{p} + E[W] + Q(E[W]) \times \frac{1}{q}}{RTT \times \left(\frac{b}{2} \times E[W] + 1\right) + Q(E[W]) \times RTO \times \frac{f(p)}{q}} \qquad \text{EQ 1–12}$$

Otherwise:

$$\gamma = \frac{\frac{q}{p} + W_{max} + Q(W_{max}) \times \frac{1}{q}}{RTT \times \left(\frac{b}{8} W_{max} + \frac{q}{pW_{max}} + 2\right) + Q(W_{max}) \times RTO \times \frac{f(p)}{1-p}} \qquad \text{EQ 1–13}$$

where:

$$E(W) = \frac{2+b}{3b} \times \sqrt{\frac{8q}{3bp} + \left(\frac{2+b}{3b}\right)^2} \qquad \text{EQ 1–14}$$

$$Q(W) = min\left(1, \frac{(1-q^3) \times (1+q^3 \times (1-q^{W-3}))}{1-q^W}\right) \qquad \text{EQ 1–15}$$

$$f(p) = 1 + p + 2p^2 + 4p^3 + 8p^4 + 16p^5 + 32p^6 \qquad \text{EQ 1–16}$$

where $q = (1-p)$ and b is the number of TCP packets acknowledged by a single Ack. The default value for b is 2.

Recently it was proposed to allow changes in the TCP congestion window using ECN [B29]. ECN has been added by the introduction of the ECN field in the IP packet header. The ECN field is two bits in the type of service (ToS) field of the IP header. A TCP source that is ECN-capable transport (ECT) will advertise its capability using one of two code points of the ECN field. When a router receives a packet with ECT indicated, it will run a RED-like algorithm with the exception that the packet will be marked with congestion experienced (CE) indication instead of being dropped.

When a packet is received by the TCP destination with CE indicated, the destination will reflect back this information back to the source using ECN-Echo field in the TCP Ack packets. The source will then react to the ECN-Echo in the same way as it reacts to packet drop. For example, the source would react to the ECN-Echo by setting its cwnd to half its current size.

This ECN-based mechanism is a binary scheme in the sense that it can signal only congestion or no congestion. However, it has no ability to indicate the severity of the congestion without a longer header field. Binary schemes may lead to oscillating behavior where the network queue oscillates between the congestion and no-congestion states, especially for cases characterized by long feedback delay. Binary schemes also suffer from fairness-related issues in which flows that traverse multiple nodes will be penalized more than flows that traverse a smaller number of nodes. This phenomenon is called the *beat-down* problem and is common in binary feedback schemes.

More elaborate feedback schemes based on rate measurement and allocation are proposed in the context of the ATM available bit rate (ABR) service [B1]. An explicit rate (ER) feedback mechanism allows network nodes to run a rate allocation algorithm that allocates different rates to the different flows traversing a node. The explicit rate value is then communicated back to the source where the source starts using the assigned rate. ER-based schemes converge faster than other schemes and do not suffer from the beatdown problem at the expense of a modest increase in the overhead.

Performance Metrics

Network and application performance are assessed in an objective manner by employing a set of metrics that reflect their performance. Common performance metrics described here are throughput, delay, and loss.

Throughput

When an application injects its traffic into the network, the application traffic is usually augmented with overhead that is required for the transport of the information at the different protocol layers. This overhead includes the necessary protocol information to transport information bits between different nodes of the network and error detection fields that ensure packet integrity. Network addressing, type of service indication, and cyclic redundancy check (CRC) are examples of such overhead.

Throughput is a measure of the quantity of information delivered over a finite period of time. It is usually expressed in terms of bits/s. If I is the amount of information transmitted over an interval of length T, then the throughput of the system, γ is given by:

$$\gamma = \frac{I}{T} \qquad \textbf{EQ 1–17}$$

An important measure related to the system throughput is the system efficiency, η, which measures the fraction of network resources spent in transporting information bits. Let O be the amount of overhead bits transported over the same interval of length T. The system efficiency is given by:

$$\eta = \frac{I}{I+O} \qquad \textbf{EQ 1–18}$$

Some systems exhibit a cyclic behavior. The system goes through cycles where the system behavior is regenerated either in a statistical or deterministic sense, as shown in Figure 1–20. For those systems the long-term (steady-state) average throughput is given by:

$$E(\gamma) = \frac{E(I)}{E(T_c)} \qquad \textbf{EQ 1–19}$$

where $E(\gamma)$ is the average value of the random variable γ.

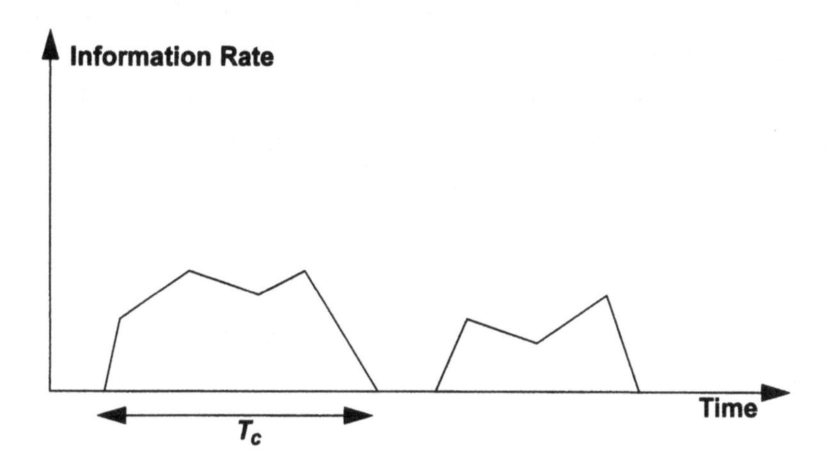

Figure 1–20: Regenerative System Behavior

Delay

In data networks, delay is measured on a packet-by-packet basis between two measuring points, MP1 and MP2, as shown in Figure 1–21. Packet transfer delay (PTD) is usually defined as the time elapsed since the arrival of the first bit of the packet at MP1 until the departure of the last bit of the same packet at MP2. The location of the two measuring points decides the scope of the delay measurements. For example, end-to-end delay is measured between two measuring points that reside at the customer equipment at the two ends of the session.

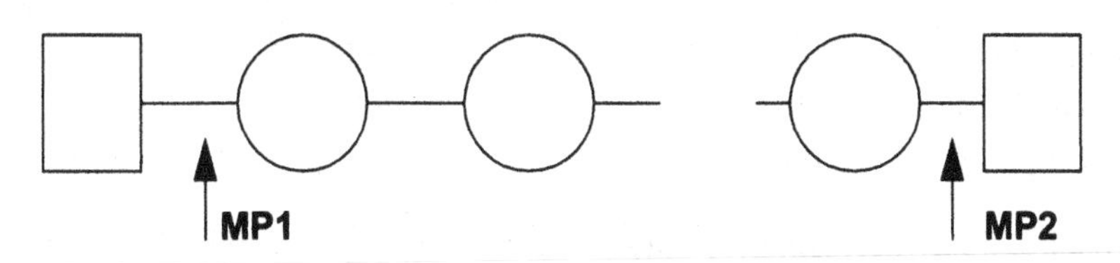

Figure 1–21: Measurement Points

Several factors contribute to the total delay. Among those factors are propagation and processing delay, medium access delay, and queuing delay. Packet delays vary from one packet to another. In general packet transfer delay is a random variable that is fully characterized by its probability density function (PDF). Figure 1–22 shows a typical packet delay PDF.

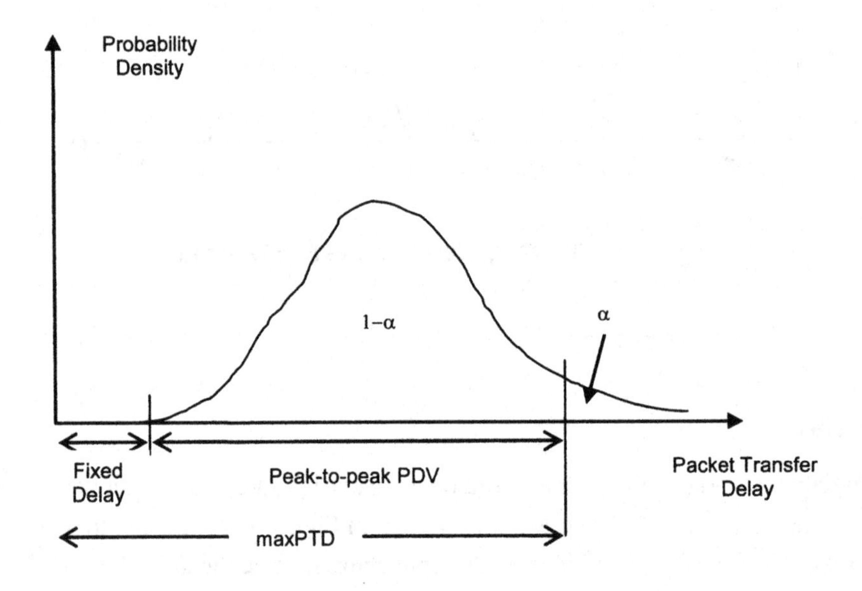

Figure 1–22: Delay Probability Density Function

Packet delay performance may be measured over the entire duration of the session by recording the delay encountered by each packet of a particular session. It can also be measured over a packet population of a finite size, e.g., over *n* packets. The results of the measuring process can be grouped to form a set of packet delays so that $\Im = \{d_1, d_2, \ldots, d_n\}$ where d_i is the measured delay of the ith packet of the population.

The packet delay performance may be measured by one or more of the following delay metrics.

Mean Packet Delay

An unbiased estimate of the mean packet delay is given by:

$$\bar{d} = \frac{1}{n}\sum_{i=1}^{n} d_i \qquad \text{EQ 1–20}$$

Maximum Packet Delay

The maximum delay is the maximum delay encountered by any packet in the selected population and is given by:

$$d_{max} = max\{d_i\} \qquad \text{EQ 1–21}$$

Packet Delay Variation

Packet delay variation is defined as the difference in the delays experienced by two packets of a given flow. The two packets are chosen according to a defined selection function [B8]. With this definition, the packet delay variation is given by:

$$\Delta_{ik} = d_i - d_k \qquad \text{EQ 1–22}$$

Packet delay variation can take positive or negative values. A positive value for the delay variation implies that the two information units are closer together at the MP_2 than at the MP_1. A negative delay value implies that they are further apart at MP_2 than at MP_1.

Delay variation is commonly measured either by:

- The difference between the maximum packet transfer delay and the mean delay observed for a particular population or
- The difference between the actual packet delay and the minimum delay value observed for a particular population

Packet Loss

Information loss can happen as a result of traffic congestion, such as buffer overflow, or because of transmission errors. The main concern is loss events due to buffer overflow. The measure of packet loss is the packet loss ratio defined as the ratio of the number of lost packets to the total number of packets transmitted.

SIMPLE QUEUEING MODELS

This section describes some of the elementary queueing models that are useful for the analysis of many systems. A later section shows how those simple models can be used to approximate the performance of more complex systems.

A queueing system is represented by an arrival process, $A(t)$, and a service process, $B(t)$, together with a service discipline [B21]. An example of a service discipline is the FIFO (First-In-First-Out) discipline, where units or customers are served in the order of their arrival. Figure 1–23 shows a realization of the arrival and the service processes where the number of units queued at any time t, $Q(t)$ is given by:

$$Q(t) = Max[0, (A(t) - B(t))] \quad \text{EQ 1–23}$$

A queueing system is usually described using the notation: $A(t)|B(t)|N|K$ where $A(t)$ and $B(t)$ are the arrival and the service process as before, N is the number of servers available, and K is the maximum number of units or customers that can be stored in the system, including those in the service process.

Two simple queueing systems will be considered here, the $M|M|1$ and the $M|G|1$ systems. The two systems differ in the probability distribution of the service time needed to complete the service of each customer. In the $M|M|1$ system, the service time is assumed to follow an exponential distribution. In the $M|G|1$ system, the exponential assumption is relaxed and the service time can have a general distribution, e.g., deterministic or uniform distribution.

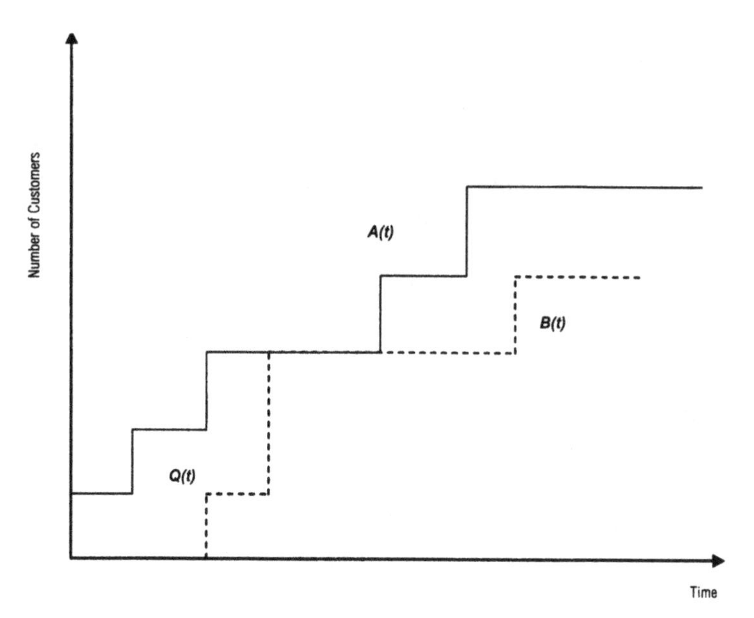

Figure 1–23: Arrival And Service Processes

In both systems the state of the queue at any given time t is represented by the number of customers (e.g., number of packets) available in the system, including the one in service. This state can be mathematically expressed as:

$$S(t) = \{Q((t) \geq 0) : Q((t) \subset Z;t)\} \qquad \textbf{EQ 1–24}$$

where Z is the set of the positive integers.

The objective is to find the steady state probability distribution for the number of customers in the system, i.e.:

$$\pi_i = Pr(Q(t) = i;(t \rightarrow \infty)) \qquad \textbf{EQ 1–25}$$

M|*M*|1 Queueing System

The $M|M|1$ queueing system is a single-server queue with infinite room (e.g., infinite buffer) to store incoming customers. It is characterized by a Poisson arrival process, which assumes that the number of arrivals, $N(T)$, over a time period of length T has a Poisson distribution of the form:

$$Pr[N(T) = n] = \frac{(\lambda T)^n}{n!} e^{-\lambda T} \qquad (n \geq 0) \qquad \textbf{EQ 1–26}$$

With the Poisson arrival process, the time t between two arrivals is known to have an exponential distribution of the form:

$$Pr[t \leq \tau] = 1 - e^{-\lambda\tau} \qquad (\tau \geq 0) \qquad \textbf{EQ 1–27}$$

where λ is the arrival rate and is expressed in terms of customers per second (e.g., packets/second).

Customers arriving at the queueing system bring an amount of work that is represented by the service time required to complete its service and leave the system. In $M|M|1$ queueing system, the service time b is assumed to have an exponential distribution of the form:

$$Pr[b \leq \tau] = 1 - e^{-\mu\tau} \qquad (\tau \geq 0) \qquad \textbf{EQ 1–28}$$

where μ is the service rate and μ^{-1} is the mean service time.

With $M|M|1$ assumptions, the system state at any time t is modeled by a Markov process and its steady state probabilities π_i exist only if $\rho = \lambda/\mu < 1$. The condition $\rho < 1$ states that for the $M|M|1$ system to be stable, the number of arrivals during an average service time has to be less than one. This condition is intuitive because at most one customer leaves every single service time. Thus for the system to be able to keep a bounded queue size on average, less than one customer must arrive for every departing customer.

Assuming $\rho < 1$, the steady state probability that the system is in state i is given by:

$$\pi_i = (1-\rho) \times \rho^i \qquad (i \geq 0) \qquad \text{EQ 1–29}$$

It is now possible to find a number of key statistics such as the average number of customers in the system, $E(Q)$, the average system time, $E(T)$, and the average waiting time in the queue, $E(W)$, so that:

$$E(Q) = \frac{\rho}{1-\rho} \qquad \text{EQ 1–30}$$

$$E(T) = \frac{1/\mu}{1-\rho} \qquad \text{EQ 1–31}$$

Note that $E(Q) = \lambda E(T)$. This relationship is called Little's formula and is applicable for other queueing systems, not just the $M|M|1$.

$$E(W) = \frac{1}{\mu} \times \frac{\rho}{1-\rho} \qquad \text{EQ 1–32}$$

Note that $E(T)$ is equal to the sum of the average time spent in the queue, $E(W)$ and the average service time, i.e.:

$$E(T) = E(W) + \frac{1}{\mu} \qquad \text{EQ 1–33}$$

Other useful variations of the $M|M|1$ systems are considered in the next section.

M|M|1|K System

$M|M|1|K$ system differs from the $M|M|1$ system in that the system has a capacity to hold a finite number of customers. The system capacity, including the customer in service, is given by K. In this case the steady state probability that there are i customers in the system is given by:

$$\pi_i = \frac{1-\rho}{1-\rho^{K+1}} \times \rho^i \quad (0 \le i \le K) \qquad \text{EQ 1–34}$$

Note that in $M|M|1|K$ the parameter ρ is allowed to exceed a value of 1 because the limited storage always keeps the queue size finite. The performance of the system is usually given by its blocking probability defined as the probability that the incoming customer will be dropped because of lack of space. This probability is given by:

$$\pi_K = \frac{1-\rho}{1-\rho^{K+1}} \times \rho^K \qquad \text{EQ 1–35}$$

M|M|C System

Similar to the $M|M|1$ system, the $M|M|C$ system has infinite room to hold customers. However, it has more than one server ready to serve customers in the queue. In this case the condition $\rho = l/c\mu < 1$ must be satisfied to ensure finite queue size. The steady state probability of having i customers in the system is given by:

$$\pi_i = \begin{cases} \pi_0 \times \dfrac{(C\rho)^i}{i!} & (i \le C) \\ \pi_0 \times \dfrac{(\rho)^i C^C}{C!} & (i \ge C) \end{cases} \qquad \text{EQ 1–36}$$

where π_0 is given by:

$$\pi_0 = \left[\sum_{k=0}^{C-1} \frac{(C\rho)^k}{k!} + \left(\frac{(C\rho)^C}{C!}\right) \times \left(\frac{1}{1-\rho}\right) \right]^{-1} \qquad \text{EQ 1–37}$$

The $M|M|C$ queuing model is useful in modeling communication systems with multiple servers such as an FDM or TDM system, where each frequency band or time slot is represented as a separate server.

M|M|C|C System

This system has C servers and the ability to only hold those customers that are already in the service process, i.e., the system has no buffer. No additional customers can be admitted to the system if there are C customers in the system each occupying one of the available servers. This system is usually referred to as *blocked customers cleared* to indicate that if no free server is available, the incoming customer is blocked or cleared from the system.

The performance of the $M|M|C|C$ is given by the probability that all of the C servers are busy and an incoming customer will be cleared. For $M|M|C|C$ system the blocking probability is given by:

$$\pi_C = \frac{\rho^C / C!}{\sum_{k=0}^{C} \frac{\rho^k}{k!}} \qquad \textbf{EQ 1–38}$$

Equation 1–38 is known as the *Erlang-B* formula and has been for many years the foundation for engineering circuit-switched networks such as the PSTN.

M|G|1 Queueing System

Unlike the $M|M|1$ queueing system, the $M|G|1$ system can not be modeled by a Markov process at any point in time. Instead, a technique called *embedded Markov process* is used for the $M|G|1$ analysis. The embedded Markov process technique relies on the fact that the $M|G|1$ system exhibit Markovian property at customer departure instants, so that

$$Q_{n+1} = Q_n - \Delta_{q_n} + v_{n+1} \qquad \textbf{EQ 1–39}$$

where Q_n is the number of customers left behind by the nth departing customer, v_n is the number of customers arriving during the nth customer service time, and Δ_{q_n} is an indicator function to indicate whether the *n*th departing customer left behind a nonempty queue.

The state of the system at the departure instance is represented by the number of customers left behind by the $n + 1$ departing customer. This state is equal to the number of customers left behind by the nth departing customer plus the number of arrivals, v_{n+1} during the service time of the $n + 1$ customer minus the customer that has just departed, if any. The average queue size $E(Q)$ can be obtained by taking the expectation of the two sides of Equation 1–39 [B21] to yield:

$$E(Q) = \rho + \frac{\lambda^2 E[x^2]}{2(1-\rho)} \quad \textbf{EQ 1–40}$$

where $\rho = \lambda/\mu$, $E[x^2]$ is the second moment of the service time, and μ^{-1} is the average service time.

Applying Little's formula finds the average time spent in the system, $E(T)$, and is given by:

$$E(T) = \frac{E(N)}{\lambda} = \frac{1}{\mu} + \frac{\lambda E[x^2]}{2(1-\rho)} \quad \textbf{EQ 1–41}$$

Example: For an $M|D|1$ system where the service time is deterministic (i.e., the time needed to transmit a packet or serve a customer is constant and is set at d seconds), the average number of packets in the queue is given by:

$$E(Q) = \rho + \frac{\lambda^2 \times d^2}{2(1-\rho)} \quad \textbf{EQ 1–42}$$

Note that the average queue size for a $M|D|1$ queue is less than that of a $M|M|1$ queue with the same utilization factor ρ. The shorter average queue size is attributed to lack of service variability when the service time has a constant value.

Chapter 2 QoS Architectures

This chapter describes some well-known and popular QoS architectures. The choice of which architecture to consider is dependent on network mission, environment, and the nature of performance assurances required. The choice of a particular QoS architecture could influence deployment of the various traffic management mechanisms described in the previous chapter.

Two QoS architectures are commonly deployed and have been the focus of the networking community for the past two decades. Those architectures are reservation-based architecture and differentiation-based architecture. A description of the main features of these two architectures and some practical examples are discussed in this chapter.

Reservation Model

The reservation model is perhaps the most familiar. It has been deployed for decades in PSTN networks. The salient feature of this model is that network resources are reserved per flow or per connection. Network resources managed on a reservation basis ensure that new flows are not admitted to the network unless there are sufficient resources in the network to accommodate them and satisfy their performance requirements. Once reserved, the resources are expected to be there for the flow to use.

The nature of the reservation itself differs by the kind of switching technology deployed by the network. For circuit-switched, connection-oriented (CS-CO) networks, resource reservation amounts to reserving a physical circuit of the appropriate bandwidth on all of the transmission facilities from the source to the destination. This practice represents a *hard* reservation, in which circuits are reserved for the exclusive use by a flow or a connection not independent of its activity. The circuit will remain unused if no traffic is generated by this particular flow. Hence hard reservation, as in the case of PSTN networks, is not suitable for bursty traffic sources that are characterized by long idle intervals. Network resources are wasted during these periods where the source is idle.

In packet-switched, connection-oriented (PS-CO) networks, resources are reserved. However, because of the sharing nature of packet switching networks, resources can be used by other connections sharing the same transmission facility when the resources are not in use by their associated flow. This setup represents *soft* reservation and is usually performed within some bounds to statistically ensure application performance. A reservation model usually requires a signaling procedure for connection establishment and release. Signaling System 7 (SS 7) has been in use for many years for signaling and resource reservation in PSTN. The signaling protocol in its abstract form can include the following messages:

- **Setup Request Message**: Usually generated by the sender (source). It indicates the sender's intent to establish a connection with a particular destination. The setup message includes elements needed for the successful establishment of a connection, such as source and destination identifications, e.g., source and destination addresses. It also includes the requested connection attributes including those related to QoS support.
- **Setup Response Message**: Sent by the destination to inform the sender of the outcome of the setup request. The message can also be generated by any of the intermediate network nodes if the connection request fails at that node for a given reason, e.g., lack of resources.
- **Disconnect Message**: Sent by the source or the destination to indicate its intent to terminate connectivity.

The signaling messages convey information related to flow identification, resources requested, resources granted, and performance expectations. The signaling messages are also used to establish connection context or a connection state at each node along the end-to-end path of the connection. The connection state includes a connection identifier and other connection attributes such as priority, resources needed, etc.

Reservation models are usually augmented with *service types*. In the PSTN the network can offer only one service, namely voice service. With multimedia applications, the need to support different service types with different attributes becomes essential. This need for multiple service types gives rise to the introduction of constant bit rate (CBR), variable bit rate (VBR) services, etc. [B2]

Reservation models are necessary for supporting absolute QoS. Absolute QoS ensures that application performance is assured either statistically or by establishing a deterministic upper bound on performance metrics. Assured (or guaranteed) performance requires the availability of resources whenever needed, and hence the need for reservation. Absolute QoS is suitable for applications that may suffer severe performance degradation unless their performance requirements are assured. Real-time applications such as voice over IP (VoIP) or IP television (IPTV) require assured delay performance to ensure timely delivery.

Reservation Model Example

Prominent examples of reservation model are those related to ATM QoS [B2] and the IP Integrated Services model [B31]. In this section we limit our discussion to the Integrated Service model. Interested readers can refer to AF-TM-0121.000 [B2] and AF-PNNI-0055.002 [B1] for detailed descriptions of ATM QoS and the associated signaling and routing.

The Integrated Service (IntServ) model was developed at the IETF to enable IP networks to support assured QoS. It includes a signaling protocol specification and service type definitions.

Signaling Protocol

The integrated service uses the Resource Reservation Protocol (RSVP) for signaling the need to establish a session with its desired attributes [B5]. Session attributes can include those related to QoS parameters, such as traffic characteristics and performance requirements.

RSVP makes use of existing routing protocols, e.g., OSPF (Open Shortest Path First), for session routing. It is then used to ensure that sufficient resources exist along the selected route to satisfy the performance requirements of the session. No attempt is made by RSVP or the routing protocol to find the "best" route that can accommodate the current request. The best route may be defined as the one with the least utilization at the instant when the session is established.

For signaling protocol, a session is uniquely identified by a combination of source address, destination address, transport layer protocol type, and destination port number. The union of these fields acts as the connection identifier.

RSVP is a receiver-controlled signaling protocol in the sense that the receiver is responsible for indicating the level of QoS needed. The reservation level is then communicated back to the source. Receiver control is a requirement to be able to support multicast.

RSVP employs two main signaling messages for session establishment. Those messages are the PATH message and the RESV message [B5]. PATH and RESV messages are transmitted in the forward and the backward directions, respectively, as shown in Figure 2–1. The PATH message is generated by the source end system and advances along the route that was independently chosen by the routing protocol. The PATH message establishes a session (connection) state in network elements, e.g., routers along the session path.

The destination end system responds to the PATH message with a RESV message. The RESV message includes the required reservation level to support the QoS level requested. The PATH message propagates in the reverse direction along the same route followed by the PATH message. Network elements receiving the RESV message perform the reservation function. If resources are not available at any one router along the selected route, session establishment fails. Failure of the session establishment is communicated back to all the network elements along the route including the end system and results in the removal of the session state.

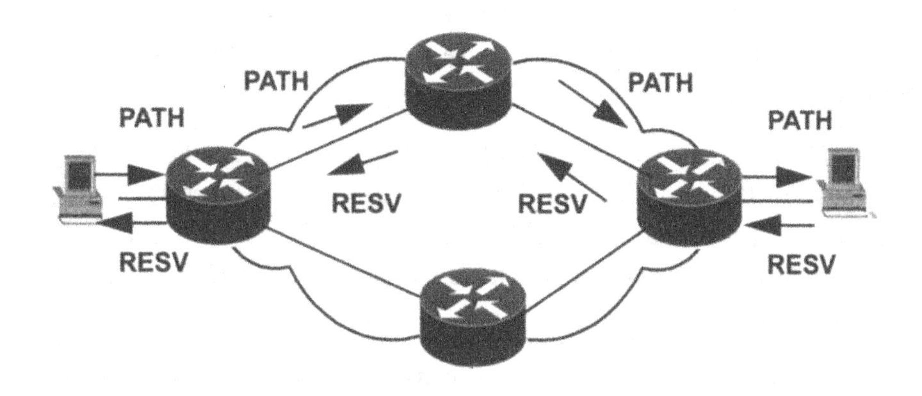

Figure 2–1: RSVP Reservation Procedure

If resources are available everywhere along the selected route, the session is successfully established, and the source end system starts sending application traffic with the expectation that resources are available to satisfy the required performance objectives. It has to be mentioned that RSVP signaling protocol reserves resources in one direction only. A REVConfirm message (not shown in the figure) can then be sent from the source to the destination to confirm the reservation.

RSVP is an example of a *soft* state signaling protocol. A session state will be maintained at a router as long as it is refreshed periodically. The refresh period is determined by a configured cleanup timeout that is used to update the reservation tables and purge old states that are no longer in use. Session state will be removed from the router unless it receives a PATH or RESV message periodically before a session timer expires. States may also be deleted explicitly by sending a teardown message. The teardown message is used to delete the current reservation and release resources and session state. It can be initiated by either the sender or the receiver.

Reservation time out is in contrast to *hard* state signaling protocols where a state always exists without the need to periodically refresh. It can only be deleted by an explicit delete or release message. A hard-state signaling protocol example is the Private Network-to-Network Interface (PNNI) used for ATM signaling [B1].

The PATH message includes fields required for source and destination identification as well as those related to the QoS request. A PATH message is required to carry a Sender TSPEC that defines the traffic characteristics of the flow that the sender will generate. The format of the Sender TSPEC is shown in Figure 2–2 [B34].

Sender TSPEC requires that sender traffic parameters (r, b) be expressed in terms of a token bucket algorithm that provides an upper bound on the sender expected traffic as described in the first chapter. The parameter p is the peak rate. In many cases it is limited by the physical rate of the line connecting the sender to the network edge. Two other parameters are included to describe the minimum and the maximum allowed packet length and are denoted as m and M, respectively. Data units that are shorter than m will be counted by the leaky bucket as having a size equal to m.

<table>
<tr><td>Service Number</td><td>Reserved</td><td>Service Data Length</td></tr>
<tr><td>ParamID-127</td><td>Parameters Flags</td><td>Parameters Data Length</td></tr>
<tr><td colspan="3">Token Rate [r]</td></tr>
<tr><td colspan="3">Token Bucket Size [b]</td></tr>
<tr><td colspan="3">Peak Data Rate [p]</td></tr>
<tr><td colspan="3">Minimum Policed Unit [m]</td></tr>
<tr><td colspan="3">Maximum Policed Unit [M]</td></tr>
</table>

Figure 2–2: RSVP Sender TSPEC

Upon receiving the PATH message, the receiver responds with a RESV message that includes the FLOWSPEC object. The format of the FLOWSPEC depends on the service type being requested and will be described in Figure 2–5 later in this chapter. In general, the FLOWSPEC includes information sent back to the source summarizing the level of the reservation achieved by the network.

Service Types

Guaranteed Service (GS)

The guaranteed service (GS) defines a service type where the delays experienced by packets of the same session are within a prespecified deterministic bound. Furthermore, packets will not be discarded because of queue overflow as long as the flow stays within its traffic limits as characterized by the token bucket parameters (r, b) [B30]. The GS service is best suited for real-time flows that require their packets to be delivered within a specified time period, or they become out of date. Examples of those flows are those associated with voice and video applications.

The service traffic is characterized by the parameters r, b, p, m, and M. These parameters are policed at the edge of the network by employing a leaky bucket with the appropriate parameter setting. Any packet that is noncompliant with these parameters is treated on a best-effort basis. Packets shorter than m will be counted as though their length is equal to m. A flow should be rejected if its requested maximum packet size is larger than the link maximum unit.

The provision of the GS requires a firm assurance on some level of resources in intermediate routers along the path of the flow. Those resources are buffer and transmission resources and are indicated by B and R. The engineering of the GS assumes the use of an ideal fluid flow model for estimating delay bound and buffer requirements and then computes error terms by which a router along the flow path deviated from such an ideal behavior. Those error terms are similar to those introduced in the first chapter in the context of the guaranteed rate scheduler.

With the ideal fluid flow model, the delay bound and buffer requirements for a flow conforming to token bucket parameters (r, b) are given by b/R and b respectively. Taking the deviation from an ideal fluid flow model into consideration and taking into account error terms introduced by scheduler architecture at the different nodes of the path, the end-to-end delay bound is:

$$T \le \begin{cases} \dfrac{(b-M)\times(p-r)}{R(p-r)} + \dfrac{M+C_{tot}}{R} + D_{tot} \rightarrow p > R \ge r \\ \dfrac{M+C_{tot}}{R} + D_{tot} \rightarrow R \ge p \ge r \end{cases} \quad \text{EQ 2–1}$$

where R is the reservation level for this particular flow.

The expression for the end-to-end delay bound takes into account the fact that a packet has a finite size that is bounded by M. Its transmission begins when the whole packet is received and reaches the head of the queue.

C_{tot} and D_{tot} are given by $C_{tot} = \Sigma C_i$ and $D_{tot} = \Sigma D_i$, respectively. C_i and D_i are the nodal maximum deviation from the fluid model. The error term C is the rate-dependent error term. It represents the delay a packet might experience due to the rate parameters of the flow. The error term D is rate independent and represents the worst case non-rate-based transit time variation through the node.

A router needs to report its local error rates to other nodes along the path of the flow to allow for the calculation of the end-to-end delay bounds. The mechanism by which this information is reported is embedded in the RSVP signaling protocol by using the defined ADSPEC object. The ADSPEC object is optional and is included in the PATH message. It includes default fields with fields that are specific to the service being requested. The general format of the ADSPEC is shown in Figure 2–3.

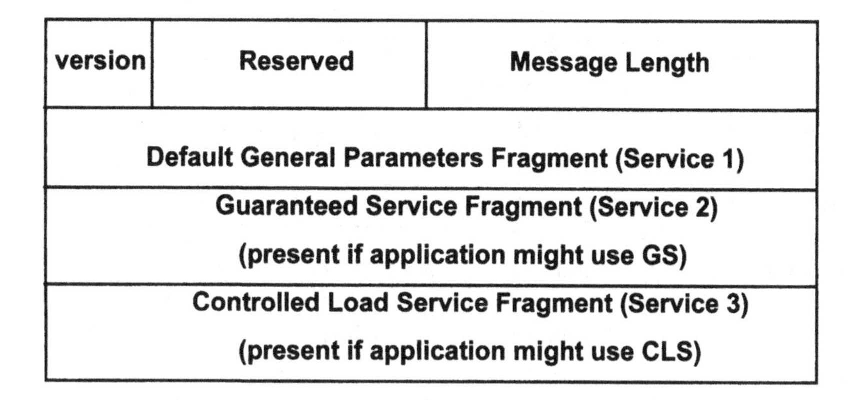

Figure 2–3: ADSPEC Information Element

The Guaranteed Service Fragments of the ADSPEC carries information needed to compute C_{tot} and D_{tot}. It is updated by the network elements along the route followed by the PATH message. The format of the Guaranteed Service Fragment is shown in Figure 2–4. Each parameter in Figure 2–4 is preceded by a Parameter ID field, Parameter Flag, and Parameter Length.

In addition to the values for the end-to-end composite C_{tot} and D_{tot}, the ADSPEC also allows for the accumulation of these values since the last point or network element where flow shaping is performed. These values are included in the C_{sum} and D_{sum} fields.

The RSVP signaling protocol uses the FLOWSPEC object to signal to the source the reservation level and the updated parameter values. The FLOWSPEC is included in the RESV message generated by the receiver. The FLOWSPEC format is shown in Figure 2–5. It includes the same parameters as in the sender TSPEC. The parameter values in the FLOWSPEC might be different from the parameter values in the TSPEC.

The FLOWSPEC specifies the level of reservation needed as indicated by the rate parameter R and the slack term S. R and S are referred to as the RSPEC.

Service Number = 2		x	Reserved	Length
Version #	Reserved	IS Length		
Service Number	Reserved	Service Data Length		
ParamID=133	Parameters Flags	Parameters Data Length		
end-to-end composed value for *C* [C_{tot}]				
ParamID=134	Parameters Flags	Parameters Data Length		
end-to-end composed value for *D* [D_{tot}]				
ParamID=135	Parameters Flags	Parameters Data Length		
Since-last-shaped point *C* [C_{sum}]				
ParamID=136	Parameters Flags	Parameters Data Length		
Since-last-shaped point *D* [D_{sum}]				

Figure 2–4: Guaranteed Service Fragment

The slack term *S* could be used by the network nodes to adjust their local reservations in cases where the required end-to-end delay is longer than the end-to-end delay bound. This feature would allow network nodes to admit more flows that would otherwise have been rejected.

Guaranteed service and its related reservation are managed based on the worst-case delay bounds. Depending on how tight the bound is, this practice can lead to an underutilized network. The problem may be alleviated with continuous measure, whenever possible, of the end-to-end delay and reporting those values back to the nodes to adjust their level of reservation.

Version #	Reserved	Length
Service Number	Reserved	Service Data Length
ParamID-127	Parameters Flags	Parameters Data Length
Token Rate [*r*]		
Token Bucket Size [*b*]		
Peak Data Rate [*p*]		
Minimum Policed Unit [*m*]		
Maximum Policed Unit [*M*]		
ParamID-130	Parameters Flags	Parameters Data Length
Rate [*R*]		
Slack Term [*S*]		

Figure 2–5: FLOWSPEC Format

Controlled Load Service (CLS)

Controlled load service (CLS) does not provide any firm assurance of the end-to-end delay [B33]. Instead, CLS packets will encounter a lightly loaded best-effort network. The end-to-end delay should not greatly exceed the minimum fixed delay due to the fixed propagation delay and the time needed to process the received packet by a node. A very high percentage of transmitted packets will be successfully delivered by the network to the receiver. The percentage of lost packets should approximate the packet error rate due to the transmission medium.

A CLS flow is characterized by the same TSPEC parameters as is the case in GS. Those parameters (r, b, p, m, and M) are included in the sender TSPEC object in the PATH message. The flow is policed at the network edge, and nonconformant packets are treated on a best-effort basis. For instance, nonconformant packets may be tagged with high discard precedence values and would be discarded first when network congestion is encountered.

Similar to the GS, an ADSPEC object may optionally be included in the PATH message. However, no parameters specific to the CLS are included in the ADSPEC other than a break bit that is set by routers incapable of supporting CLS.

The RESV message includes a FLOWSPEC object. Similar to the case in GS, the FLOWSPEC includes TSPEC parameters that could have different values than the values transmitted in the PATH message. CLS differs from GS in that the FLOWSPEC does not include as part of its parameters the RSPEC because CLS needs no explicit reservation.

DIFFERENTIATION MODEL

A differentiated architecture for QoS is suitable for the support of relative QoS with no hard assurances on application performance [B4]. Packets belonging to different classes of traffic are treated differently from each other. For example, class *x* traffic might always be transmitted before class *y* traffic. Therefore, packets belonging to class *x* will always experience lower delay than that experienced by the class *y* packets. However, there are no assurances of the level of performance experienced by each class.

The forwarding treatment each flow will receive is indicated in a special field of the packet header. Therefore, only edge nodes are needed to keep a state for each of the flows passing through them. This is particularly beneficial for nodes that are required to support a large number of flows and may suffer from scalability issues related to the flow identification space as is the case for the reservation model.

The differentiating model best suits connectionless networks where decisions for routing the incoming packet are based on information in the packet header. It is a direct extension to the connectionless network paradigm to add in the packet header information related to the QoS treatment a packet expects in its header.

Admission control is usually not a requirement for the differentiating model. Because flow state is not available to the network elements, they can not participate in admission control based on per-flow requirements. Attempts to add admission control to the differentiating model are centered on the introduction of a central entity to manage the bandwidth in the whole network or a network segment, as shown in Figure 2–6. The central entity is usually referred to as bandwidth broker (BB) or bandwidth manager.

Ideally the BB needs to maintain an updated database of link resources and their current usage. The sender will signal to the BB its required level of resources, and the BB needs to make sure that those resources are available before granting admission. The need to maintain a database related to network resources usage can impose some scale limitation on the BB. It might also be required to divide the network into a number of smaller segments, each managed by a separate BB. Procedures for BB-to-BB communications must be specified to allow for checking of transmission resources in all segments.

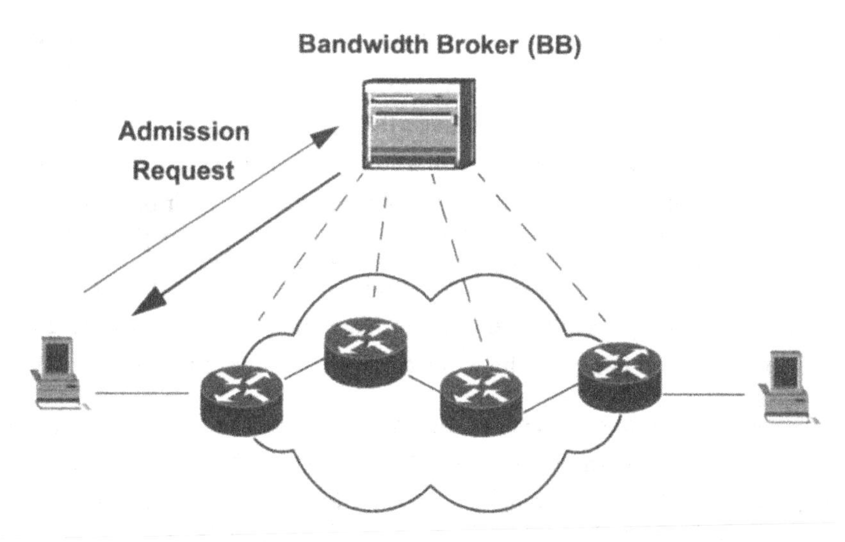

Figure 2–6: Admission Control for Differentiating Model

Differentiating Model Examples

There are two prominent examples of the differentiating models. Those are the IP-differentiating service (Diffserv) [B4] and the Ethernet user priority bits [B15].

IP Differentiated Service (Diffserv)

IP Diffserv provides a framework for QoS support on IP networks. Diffserv was introduced mainly to deal with the perceived scaling issues related to the IntServ model and its related signaling using RSVP. Those scalability issues are related to the need for the network elements to maintain per flow state and the need to periodically perform state refresh by exchanging RSVP, PATH, or RESV messages.

Diffserv architecture requires flow state to be maintained at the network edge nodes (boundary nodes) only. Network interior nodes are not required to maintain any per-flow state. Instead, interior nodes operate on a behavior aggregate (BA) flow where incoming packets are directed to the appropriate transmission queues based on information carried by the IP packet header as defined by the Differentiated Service Code Points (DSCP).

The Diffserv model defines actions needed both at the boundary and at the interior nodes [B4] to support network QoS. At the edge nodes, incoming packets are classified according to some criteria or predefined policies. Classification is usually performed based on a subset of the packet header fields. For example, incoming packets can be classified based on their behavior aggregate so that all packets are classified based on their DSCP. Packets that have the same DSCP marking will have similar treatment. Alternatively, the edge node can employ a multifield (MF) traffic classifier where incoming IP packets are classified based on the value of one or more header fields.

A traffic stream selected by the classifier will be subject to a traffic conditioning at the edge node, as shown in Figure 1–3. Each classified flow will have a traffic profile that may be defined using token bucket parameters (r, b) as described in the first chapter (see "Leaky Bucket" on page 5). Flows identified by the classifier are metered based on the associated traffic profile. The outcome of the metering function is a set of compliant and a set of noncompliant packets relative to the traffic profile. The marking function sets the DSCP accordingly. It is also possible to use the dropper to drop packets that are deemed nonconforming. A shaper might also be supported to smooth the arrival of packets to downstream nodes, if necessary.

Marking at the edge sets the DSCP field of the IP packet header to indicate a particular Per-Hop Behavior (PHB). In IPv4, the DSCP field is the first six bits of the IP ToS byte, as shown in Figure 2–7. The last two bits of the ToS field are used for the explicit congestion notification (ECN) operation and are not applicable to IP Diffserv. PHB indicates the forwarding treatment a packet expects to receive as it traverses the different network nodes.

While it is possible to define up to 64 (2^6) PHB, only a limited set of PHB is defined by the IETF. Those behaviors are the Class Selector (CS), The Expedited Forwarding (EF), and the Assured Forwarding (AF), in addition to the Default Forwarding (DF), which supports the IP legacy best-effort service.

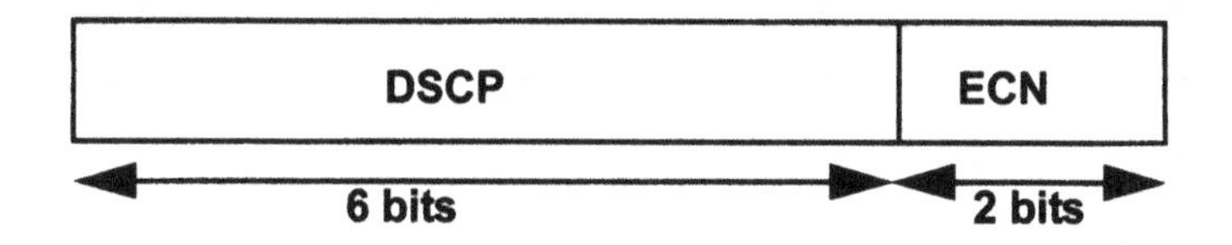

Figure 2–7: IP Type of Service (ToS) byte

The Class Selector (CS) PHB

The CS PHB group was introduced [B26] to provide a level of backward compatibility with the IP precedence bits (the first three bits of the ToS field). The CS PHB group defines up to eight classes, which are indicated by code points of the form *xxx* 000, where *x* can be 0 or 1. The CS PHB group must satisfy the following set of requirements:

- The CS code points must yield at least two independently forwarded classes of traffic.
- Packets marked with CS code point with higher numerical value (e.g., 111 000) must have a better probability of timely forwarding than packets marked with lower numerical value (e.g., 000 000).
- Packets marked with code points 11*x* 000 must be given preferential treatment over packets marked with code point 000 000.

Among the classes defined by the CS PHB group is the default forwarding (DF) that is used to support legacy best-effort applications. DF has the code point 000 000, which is the lowest priority in the CS classes.

The Expedited Forwarding (EF) PHB

The EF PHB is a forwarding treatment aimed at providing services with low loss, low delay, and low delay variation [B7]. These objectives are achieved when appropriately marked packets usually encounter short or empty queues. EF PHB is suitable for real-time applications such as voice and video. A short queue will also keep packet loss to a minimum.

A packet of length L_j arriving at time a_j at a node that supports EF PHB at a configured rate R must depart the node at time d_j so that:

$$d_j \leq f_j + \varepsilon \qquad \text{EQ 2–2}$$

where ε is the error term for the treatment of EF packets, and f_j is the ideal departure time of the jth packet from the node and is given by:

$$f_j = max(a_j, min(d_{j-1}, f_{j-1})) + \frac{L_j}{R} \qquad \text{EQ 2–3}$$

The ideal departure time of the jth packet takes into account whether the preceding packet departed before or after its ideal departure time. If it departs before its ideal departure time, then the ideal time for the jth packet to start service is the actual departure time of the preceding packet. Otherwise, the ideal time to start service for the jth packet is the ideal departure time of the preceding packet. After the packet starts transmission, the packet will take L_j/R s to depart the scheduler. The error term ε depends on the node architecture and the scheduling mechanism.

The recommended DSCP for EF PHB is 101 110.

The Assured Forwarding (AF) PHB Group

The AF PHB group defines a set of four AF classes [B13]. Within each class, an IP packet is assigned one of three levels of drop precedence. Packets with high drop precedence are more likely to be discarded than those with lower drop precedence, as explained in the previous chapter. It is often the case that when nodes are congested; they will start dropping high precedence packets first. Figure 2–8 shows a possible implementation of the AF PHB group.

The AF PHB group does not specify relative performance between the four AF classes. Packets in one AF class are forwarded independently from packets in another AF class. A node must allocate a configured minimum amount of resources to each implemented AF class. However, unlike the EF PHB, no delay requirement needs to be achieved by any of the AF classes.

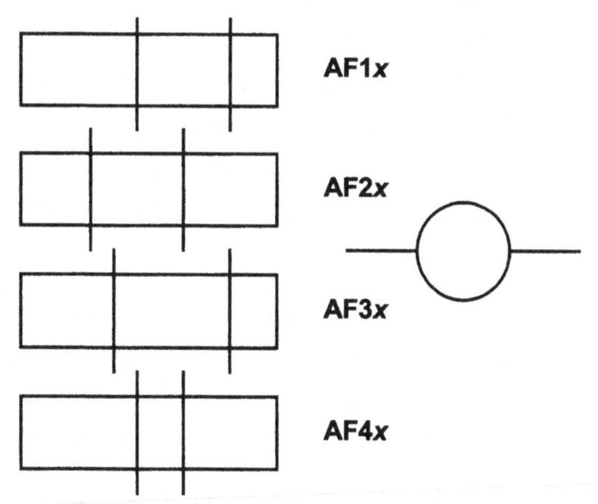

Figure 2–8: AF PHB Group

The AF PHB group is suitable for services and applications that require the transport of packets with different importance levels. The importance of the packet may be determined by the application itself, such as the case in video application where video coding produces video frames of different importance. Alternatively, packet importance can result from the conformance testing done at the edge nodes to ensure that a flow of packets adheres to its traffic parameters.

The recommended AF code points are shown in Table 2–1.

Table 2–1: AF PHB Group DSCP

	Class 1	Class 2	Class3	Class 4
Low Drop Precedence	001010	010010	011010	100010
Medium Drop Precedence	001100	010100	011100	100100
High Drop Precedence	001110	010110	011110	100110

Service Definition in Diffserv Environment

Unlike the IP IntServ, Diffserv is not provided with service types. Therefore in Diffserv the equivalent of GS and CLS service types does not exist. The decision to introduce a toolbox rather than defined service types is motivated by the need to allow the service provider to have flexibility. Service providers can create services that suit their business models instead of specifying particular services that might prove useless later and do not meet customer requirements.

Service definition in a Diffserv environment can be achieved by applying appropriate rules at the edge of the network and by applying consistent treatment at the different nodes of the network [B19]. Therefore, it is customary to write service definitions using the Diffserv model as:

Service = Edge Rules + Nodal Treatment

The edge rules are implemented at the provider edge using traffic conditioning. Rules define how to meter, mark, remark, and drop packets of the incoming flow. The nodal treatment is reflected in the PHB as signaled by the DSCP.

A possible example is shown in Figure 2–9, where the customer wanted the IntServ GS service type using the Diffserv model. Figure 2–9 shows incoming GS flows are traffic conditioned at the edge node. Each GS flow is metered according to source TSPEC parameters p, r, b, m, and M. The rule at the edge node is that any packet that does not conform to the TSPEC parameters will be marked to be forwarded as on a best-effort basis. This rule is the same as defined in GS specifications [B30]. Alternatively, a different edge rule can mandate dropping of those nonconforming packets.

All conforming packets that belong to any of the GS flows will be marked with the EF DSCP. Provider nodes will be configured with the appropriate rate R to handle the incoming GS flows. GS flows are aggregated based on their DSCP marking and have the same treatment at the provider nodes. There is no need to keep state information at these nodes.

Not shown in Figure 2–9 is the need for an admission control module to ensure that the total number of admitted GS flows does not exceed the config-

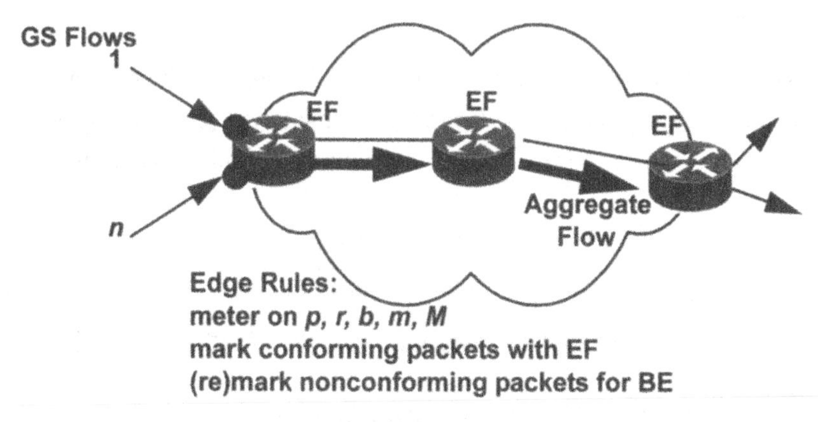

Figure 2–9: GS Type Construct Using Diffserv Model

ured EF rate *R*. The admission control module can be based on the bandwidth broker, as discussed before.

IEEE Standard 802.1Q Ethernet Frame Priority

Traditionally Ethernet has been a multicast connectionless shared medium with no QoS capabilities. With the advances in Ethernet switching and the support for Ethernet virtual LAN (VLAN) capabilities, QoS support has been added to the Ethernet bridging functionality [B15].

QoS introduction to Ethernet is in the form of three bits in the Ethernet frame header that define the priority of the frame. Therefore, up to eight priority levels are supported. The three bits are called user priority (UP) bits and are included in the tag control information (TCI) in the tagged Ethernet frame header, as shown in Figure 2–10. Other fields in the TCI are the 12-bit VLAN Identifier (VID) field and the Canonical Format Indicator (CFI) bit. The VID field identifies the virtual LAN to which the tagged Ethernet frame belongs. CFI bit is set depending on the media access control method [B15].

The frame priority (differentiation) described previously based on the UP bits is, in many aspects, similar to the support of the Class Selector PHB in the IP-differentiated service. Implementation of the UP bits can easily be supported by a strict priority scheduler with up to eight priority levels. The IEEE Standard 802.1Q specification also allows for different scheduler implementations such as RR or WRR.

When first introduced, the UP bits supported transmission priority only with no support for discard precedence. With the introduction of provider backbone bridges and the need for more elaborate support of both transmission

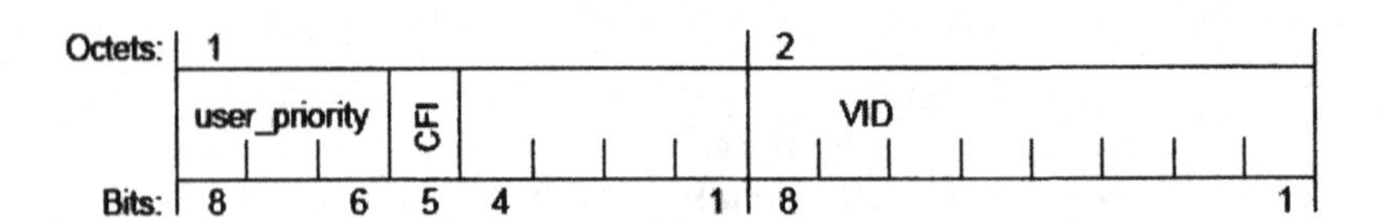

Figure 2–10: Tag Control Field

and discard priorities, the use of the three UP bits has been modified [B14] to support both these priorities.

With the new definition it is now possible to assign some of the eight priority code points (PCP) to support discard precedence. The allowed variations are given in Table 2–2.

Table 2–2: Support of Discard Eligibility in Ethernet

PCP	7	6	5	4	3	2	1	0
8P0D	7	6	5	4	3	2	1	0
7P1D	7	6	4	4DE	3	2	1	0
6P2D	7	6	4	4DE	2	2DE	1	0
5P3D	7	6	4	4DE	2	2DE	0	0DE

Table 2–2 shows the allowed mix of transmission priorities and discard eligibility. The default is the case where eight transmission priorities are supported with no support for discard eligibility. The seven P1D case supports seven transmission priorities and one discard eligibility. The discard precedence is indicated by PCP = 5 and PCP = 4, where the two PCPs are mapped to the same transmission queue, but frames marked with PCP 4 (100) are interpreted as having a higher discard precedence (or discard eligibility) than frames marked with PCP 5. The same explanation is true for 6PD2 and 5P3D.

Figure 2–11 shows an example where 5P3D is supported by a scheduler implementation. Figure 2–11 shows five transmission priority queues. Three of them are supporting two discard priority levels by implementing a discard threshold on each of them.

The change in the semantic of the UP bits is part of the IEEE 802.1 provider bridge specification [B14]. These extensions are only relevant to the carrier network. The customer network is still required to use the UP bits, as originally defined.

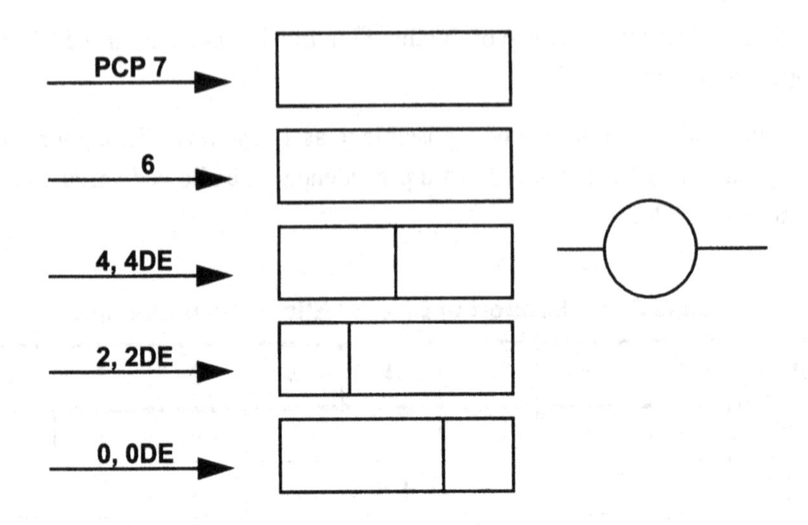

Figure 2–11: 5x3 Ethernet Differentiated Service

With the enhancement of the use of the UP bits, all the ingredients are available to support Ethernet-differentiated services similar to those of the IP networks. In Figure 2–11 both Ethernet-expedited forwarding (E-EF) and Ethernet-assured forwarding (E-AF) can be defined in the same way they are defined for IP. Unlike the case with IP Diffserv, the E-AF in this case will have two levels of discard precedence instead of three. These similarities allow for smooth interworking between Ethernet UP and IP Diffserv.

With the introduction of UP Ethernet, bridged networks are able to support real-time applications such as VoIP and non-real-time applications such as file transfer. Real-time application traffic is expected to be mapped to a higher UP value that takes transmission precedence over the UP to which non-real-time application traffic is mapped. This mapping ensures that real-time traffic is always selected for transmission before other traffic. Furthermore, data frames can be classified based on the discard precedence by using one of the possible discard eligibility methods from Table 2–2. The use of discard eligibility will allow data frames to be marked based on conformance testing, as explained in Chapter 1.

Chapter 3 IEEE Standard 802.11 Overview

This chapter provides an overview of the IEEE 802.11 architecture with emphasize on its basic MAC and its operation. A brief summary of the different IEEE 802.11 physical layers (PHY) is also included.[1]

The purpose of this chapter is to provide the reader with sufficient background in IEEE 802.11 technology to allow a better understanding and appreciation of the WLAN traffic management features to be introduced in the next chapter. In effect this chapter describes the various MAC functionality of "legacy" IEEE 802.11 equipment. The word *legacy* here refers to WLAN equipment that does not support QoS and traffic management features. The description presented is based on IEEE Standard 802.11 [B16].

IEEE 802.11 Architecture and Frame Format

The main architectural component of IEEE Standard 802.11 is the basic service set (BSS). The IEEE 802.11 BSS is designed to operate in one of two possible modes of operations, the *ad hoc* network and the *infrastructure* network.

In an ad hoc network, mobile units (MU) or stations (STA) communicate with each other in a peer-to-peer relationship. No preplanning is involved. This arrangement is usually referred to as the independent BSS (IBSS).

In an infrastructure mode, stations communicate with each other through an entity called the access point (AP). Figure 3–1 shows an IEEE 802.11 architecture that includes two BSS. STA belonging to different BSS can still communicate together using the service of the distribution system (DS) to form the extended service set (ESS). The wireless medium (WM) and the DS

---

[1] For a more detailed overview of the physical layer and MAC layer, refer to the *IEEE 802.11 Handbook: A Designer's Companion*, 2nd Edition, by Bob O'Hara and Al Petrick, ISBN-0-7381-4449-5

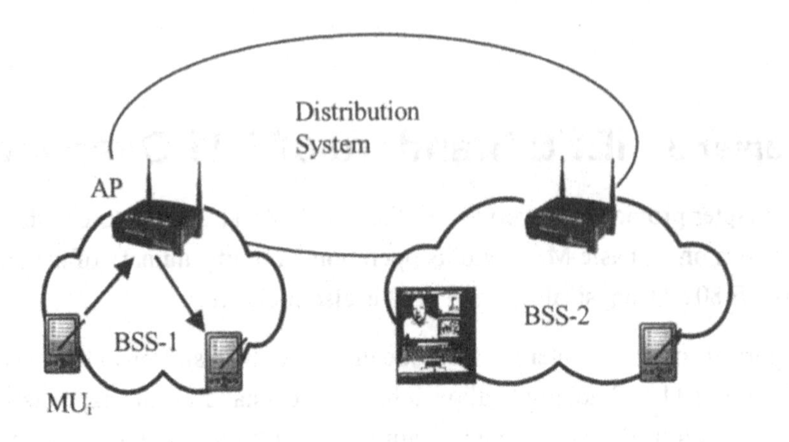

Figure 3–1: Basic Service Set (BSS)

medium (DSM) are logically separate. The DS can be implemented either wirelessly or wireline, using, for example, wireline Ethernet.

To facilitate data exchange by the DS, the IEEE 802.11 frame formats use more than the usual two MAC addresses, source address (SA) and destination address (DA), commonly used in IEEE 802.3 frames. IEEE Standard 802.11 has three frame types, the data frames, the control frames, and the management frames. Within each frame type, a number of subtypes are defined. The general IEEE 802.11 frame format is shown in Figure 3–2.

The first field of the general frame format is called the Frame Control field. As shown in Figure 3–3. The frame control field includes information related to the frame type and its subtype. The different frame types and subtypes are given in Table 3–5 on page 90.

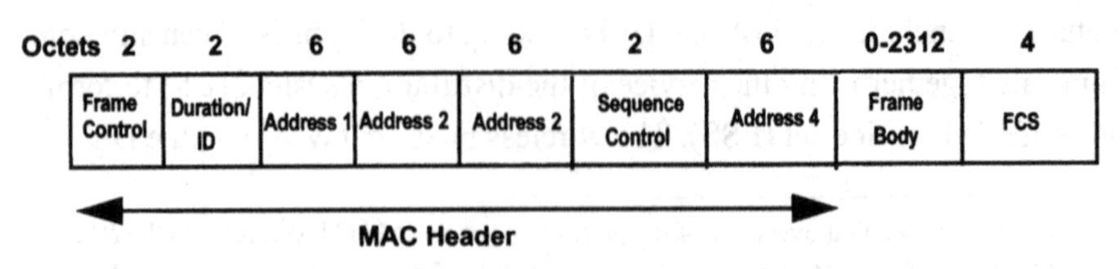

Figure 3–2: IEEE 802.11 General Frame Format

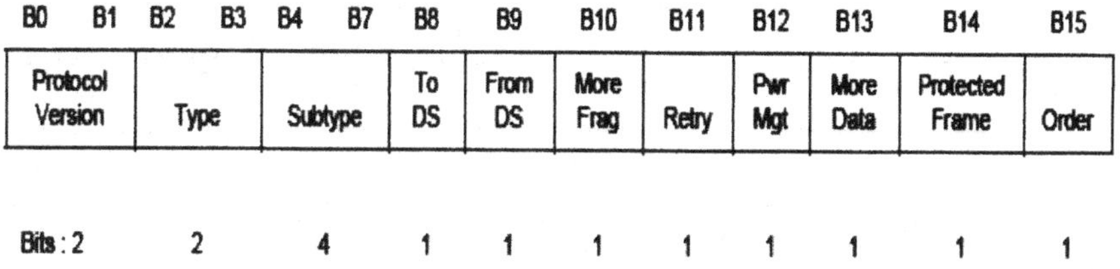

B0 B1	B2 B3	B4 B7	B8	B9	B10	B11	B12	B13	B14	B15
Protocol Version	Type	Subtype	To DS	From DS	More Frag	Retry	Pwr Mgt	More Data	Protected Frame	Order
Bits : 2	2	4	1	1	1	1	1	1	1	1

Figure 3–3: Frame Control Field

The To DS and From DS values are used in relation to the four-address format to coordinate the source and the destination of data frames, as shown in Table 3–1.

Table 3–1: To/From DS Meaning

To DS and From DS Values	Meaning
To DS = 0 From DS = 0	A data frame direct from one station to another within the same BSS or IBSS, as well as management and control frames.
To DS = 1 From DS = 0	A data frame destined to the DS or being sent by a station associated with an AP to the port Access Entity in that AP.
To DS = 0 From DS = 1	A data frame exiting the DS or being sent by the port access Port Access Entity in an AP.
To DS = 1 From DS = 1	A data frame using the four-address format. This use is not defined in IEEE Standard 802.11.

In an infrastructure mode, a STA associates with an AP using the Association Request management frame. The Association Request frame format is shown in Figure 3–4 and includes information elements (IE) related to the STA capabilities, security, and the service set identification (SSID).

Capabilities	Listen Interval	SSID	Supported Rates	Extended Supported Rates	Power Capability	Supported Channels	RSN

Figure 3–4: Association Request Management Frame

The AP responds to an Association Request with an Association Response management frame. The Association Response frame format is shown in Figure 3–5. It includes IE related to the result of the association process as reflected in the status code, the supported rates, and the AID.

Capabilities	Status Code	AID	Supported Rates	Extended Supported Rates

Figure 3–5: Association Response Management Frame

The AP periodically generates Beacon management frames to coordinate the operation of the BSS and to distribute synchronization information. In an infrastructure mode, the AP defines the timing for the entire BSS by scheduling a Beacon frame as the next transmission at regular time interval called TBTT (Target Beacon Transmission Time). The transmission of Beacon frames may still be delayed beyond the scheduled TBTT because of other transmissions taking place at the BSS.

When a STA in the BSS receives a valid Beacon, it adjusts its TSF (Time Synchronization Function) value to the current time as computed by the value of the time stamp in the received Beacon and an offset value computed at the receiving station.

IEEE 802.11 PHY Overview

The support of WLAN QoS requires a PHY layer design capable of overcoming the various interference encountered by the wireless medium. This requirement is particularly important for WLAN PHY design because it operates in unlicensed bands where interference from other devices such as microwave ovens is common.

As of the time this book was written, four PHY layers had been defined in IEEE 802.11 standards. More are expected in the future to allow the support of higher data rates. The four PHY layer standards are IEEE 802.11 PHY, IEEE 802.11a, IEEE 802.11b, and IEEE 802.11g. The IEEE 802.11 PHY was the first introduced in the 1999 edition of the IEEE 802.11 standards and supported a data rate up to 2 Mbps. It was followed by the IEEE 802.11b standards with data rates up to 11 Mbps. Then came the IEEE 802.11a and the IEEE 802.11g standards with supported rates up to 54 Mbps.

This section describes briefly the three commonly used PHY layers 802.11a/b/g. It is worth noting that the IEEE 802.11 MAC runs on top of the PHY layers almost unchanged. The relationship between the IEEE 802.11 MAC and PHY is shown in Figure 3–6.

IEEE 802.11a PHY

IEEE Standard 802.11a operates in the ISM (industrial, scientific, and medical) 5.725-5.925 GHz unlicensed band, and it supports rates up to 54 Mbps.

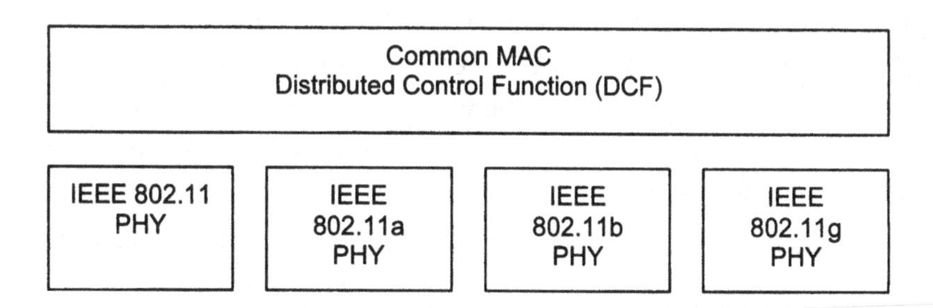

Figure 3–6: IEEE 802.11 Common MAC

The IEEE 802.11a PHY makes use of the orthogonal frequency division multiplexing (OFDM). With OFDM the available channel bandwidth is divided into a number, K, of equally spaced subchannels and a subcarrier is associated with each subchannel. The subcarrier frequencies f_i are chosen so that each subcarrier has an integer number of cycles within a given interval T, and each adjacent subcarrier differs by one cycle, i.e.:

$$f_k - f_i = \frac{n}{T} \qquad \textbf{EQ 3–1}$$

for $n = 1,2,...$ This property ensures that the OFDM subcarriers are orthogonal. T is called the symbol duration, and it is K times longer than the case where a single carrier is used. Therefore, OFDM is efficient for handling intersymbol interference (ISI) because the OFDM symbol duration is likely longer than the channel time dispersion.

Each subcarrier is then data modulated using Quadrature Amplitude Modulation (QAM) or Phase Shift Keying (PSK). In IEEE 802.11a subcarriers are modulated using binary PSK (BPSK), quadrature PSK (QPSK), 16-QAM, or 64-QAM. IEEE Standard 802.11a supports data rates of 6, 9, 12, 18, 24, 36, 48, and 54 Mbps.

When a MAC protocol data unit (MPDU) is passed to the physical layer, the physical layer adds the physical layer convergence procedure (PLCP) header and PLCP preamble to form the PLCP protocol data unit (PPDU), as shown in Figure 3–7. It is worth noting that the SIGNAL field of the PPDU is transmitted using the most robust combination of BPSK modulation and Convolutional code with a coding rate of 1/2. A coding rate of 1/2 implies that for each information bit entering the convolutional encoder circuit, two code bits will be generated at the output of the circuit. The additional bits are sometimes referred to as parity bits and are used for recovering the original information bits at the receiver. The rest of the fields are transmitted with a data rate as specified in the RATE field.

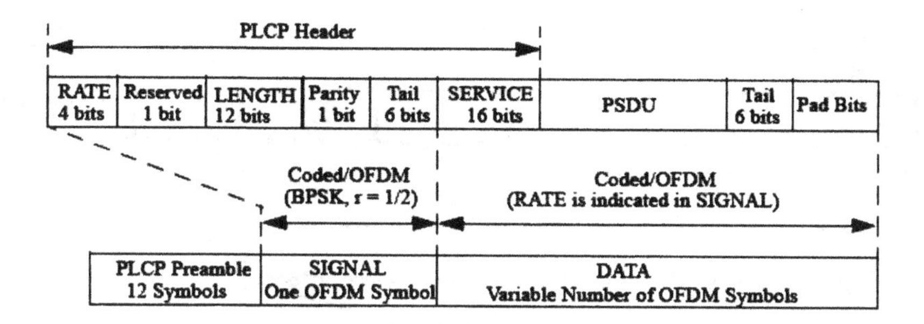

Figure 3–7: IEEE 802.11a PHY Convergence Sublayer

IEEE 802.11b PHY

IEEE Standard 802.11b operates in the ISM 2.4–2.5 GHz unlicensed band and supports data rates up to 11 Mbps. It makes use of direct sequence spread spectrum (DSSS). Spread spectrum is a means of transmission in which a signal occupies a bandwidth in excess of the minimum bandwidth necessary to send it. DSSS makes use of spreading (or chipping) sequence to distribute the signal in a higher dimension space. Spread spectrum is known to have an advantage in handling the effect of interference and/or jammers with finite power. Under these circumstances, a processing gain that is equal to the ratio of the bandwidth of the spread-spectrum signal to that of the original signal can be achieved.

IEEE Standard 802.11b is also known as high-rate DSSS (HR/DSSS). The first IEEE 802.11 PHY specification supported rates of 1 Mbps and 2 Mbps using a Baker 11-chip spreading sequence and differential binary phase shift keying (DBPSK) and differential quadrature phase shift keying (DQPSK) baseband modulation. IEEE 802.11b PHY supports higher rates of 5.5 Mbps and 11 Mbps with the use of complementary code keying (CCK) modulation.

IEEE Standard 802.11b supports long and short PPDU, as shown in Figure 3–8 and Figure 3–9. In the long PPDU, the PLCP preamble and header are always transmitted at 1 Mbps. The rest of the PPDY is transmitted at the rate specified in the signal field. In the short PPDU format, the PLCP preamble is transmitted at 1 Mbps, and the PLCP header is transmitted at 2 Mbps.

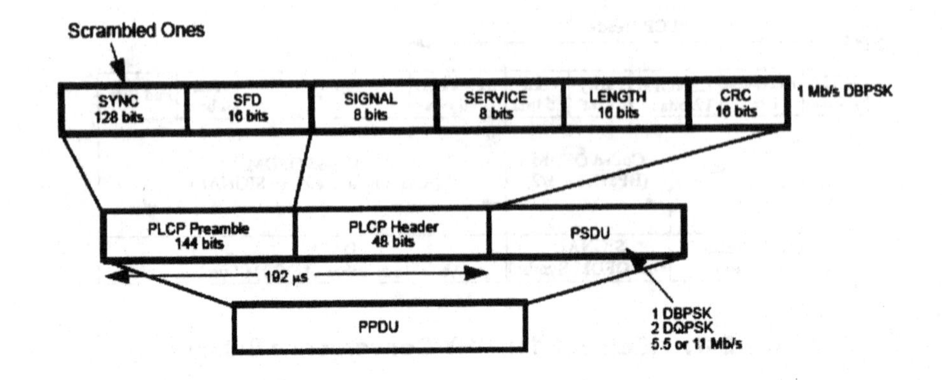

Figure 3–8: IEEE 802.11b Long PPDU Formats

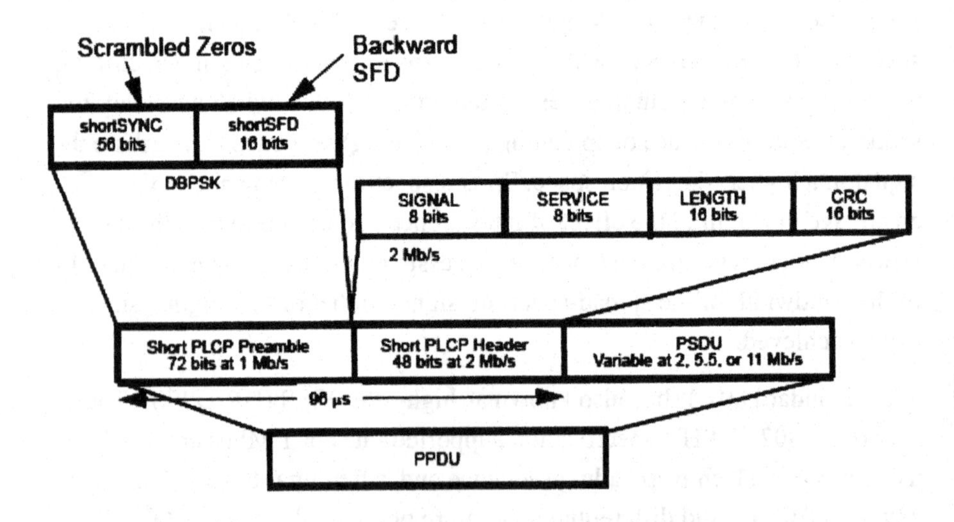

Figure 3–9: IEEE 802.11b Short PPDU Formats

The effect of the short PPDU format is to reduce the PHY overhead by almost a 100 µs with impact on system throughput, as will be discussed in "IEEE 802.11 Throughput" on page 86.

The IEEE 802.11g PHY Overview

The IEEE 802.11g PHY is also referred to as the extended-rate PHY (ERP). It operates in the ISM 2.4-2.5 GHz range. It supports multiple modes of operations that facilitates the interworking with other IEEE 802.11 PHY. The modes supported by IEEE 802.11g are:

- **ERP-DSSS/CCK**: This mode follows the same PHY recommendation as in IEEE 802.11b with a mandatory support for the short PLCP PPDU header. It supports PHY rate of 1, 2, 5, and 11 Mbps and is fully compatible with IEEE 802.11b. The support for the ERP-DSSS/CCK in IEEE 802.11g standard is mandatory.
- **ERP-OFDM**: This use the same capabilities as in IEEE 802.11a with few exceptions. It supports PHY rate of 6, 9, 12, 18, 24, 36, 48, and 54 Mbps. It is fully compatible with IEEE 802.11a. The support of ERP-OFDM in IEEE 802.11g standard is mandatory.
- **ERP-PBCC**: It is an optional mode. It is a single-carrier modulation scheme that encodes the payload using a 356-state packet binary convolutional code (PBCC). It supports payload with data rates of 22 and 33 Mbps.
- **DSSS-OFDM**: It is an optional mode of operation. It is a hybrid modulation scheme that combines a DSSS preamble and header with an OFDM payload transmission. It supports payload with data rates of 6, 9, 12, 18, 24, 36, 48, and 54 Mbps.

The IEEE 802.11g supports three types of preamble and header formats. The three types are based on the formats that are shown in Figure 3–7, Figure 3–8, and Figure 3–9 with the necessary changes to support the new modes.

The long and the short preamble support of DSSS/CCK is as described in IEEE 802.11b. They also support the ERP-PBCC and the DSSS-OFDM modes of operation at all the available payload rates.

The support of the ERP-PBCC is facilitated by the use of three bits in the SERVICE field to indicate the presence or the absence of the ERP-PBCC and reflects the length extension rules.

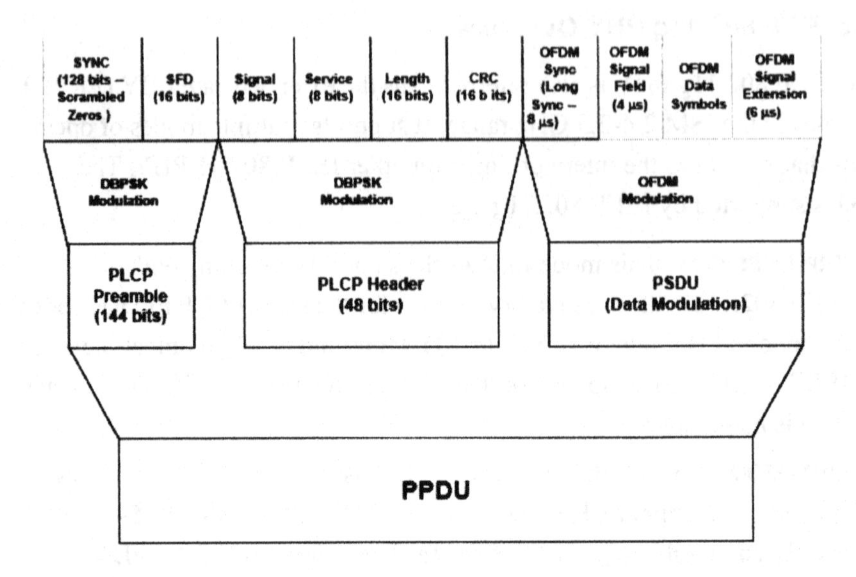

Figure 3–10: IEEE 802.11g DSSS-OFDM PPDU Formats

The long preamble PPDU format for the DSSS-OFDM is shown in Figure 3–10. The PLCP preamble and header are transmitted at 1 Mbps using DBPSK as before. The rest of the PPDU is transmitted using OFDM modulation with the addition of the OFDM extension of 6 µs.

The ERP-OFDM format, preamble, and header follows those shown in Figure 3–7. In ERP-OFDM, an ERP packet is followed by a 6-µs period of no transmission called *signal extension*, which extends the calculated transmission time by that amount. The signal extension is necessary to ensure consistence settings virtual carrier sensing parameters between the IEEE 802.11a PHY and this mode of operation of the IEEE 802.11g PHY.

The ERP-PBCC makes use of a 256-state convolutional code with a rate of 2/3. The decoder generates three bits for each two data bits entering the decoder. The three bits are then mapped to one of 8-PSK signals for transmission.

The IEEE 802.11 MAC Overview

This section describes the basic MAC features, assuming a legacy station supporting none of the traffic management features defined in Chapter 1. It is important to first understand the basic MAC functionality of a legacy station in order to pave the way for understanding those traffic management features that were introduced later.

The IEEE 802.11 MAC architecture is shown in Figure 3–11. Figure 3–11 shows two possible access functions, the Distributed Coordination Function (DCF), which is also called the *common MAC*, and the Point Coordination Function (PCF).

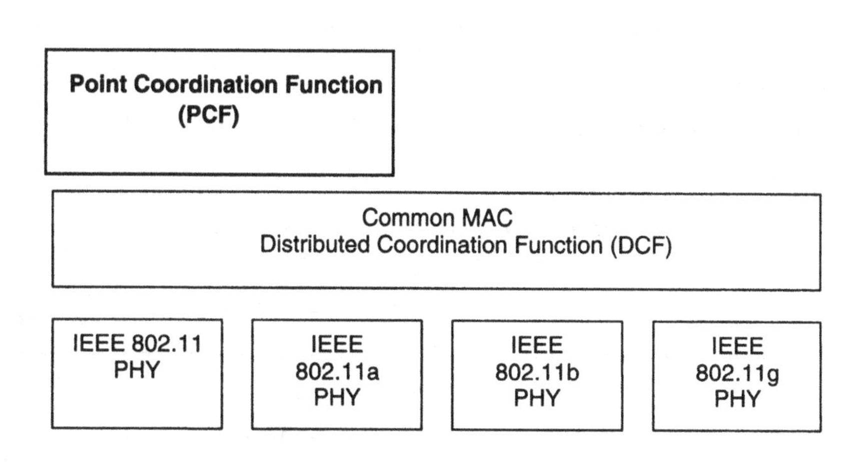

Figure 3–11: IEEE 802.11 MAC Architecture

Distributed Coordination Function (DCF)

The DCF is the basis for the shared medium access in WLAN based on IEEE Standard 802.11. It implements a form of carrier sense multiple access (CSMA) with collision avoidance (CSMA/CA). With DCF, a station contends for the medium when its buffer has frames (data or management frames) ready for transmission. The contention starts with the station sensing the medium to determine whether it is busy or idle. If the medium is idle for a period of time equal to distributed interframe space (DIFS), the station can then start transmitting a buffered frame. If the medium is busy, the station defers until the medium is determined to be idle without interruption for a period of time equal to DIFS or equal to extended interframe spacing (EIFS) when the last frame on the medium was not received correctly. After DIFS or EIFS, the station then enters the backoff state and remains there for a period of time equal to backoff_time. The backoff_time is determined randomly by:

$$\text{backoff\_time} = \text{random\_number} \times \text{aSlotTime}$$

The random_number is uniformly distributed between 0 and cw, or the *contention window*. The cw value is in the range [aCWmin, aCWmax]. Initially the value of the cw parameter is set at aCWmin.

The aSlotTime is a PHY parameter. Table 3–2 summarizes the aSlotTime parameter values for the different IEEE 802.11 PHYs described in this chapter.

Table 3–2: PHY Parameter Values

		aSlotTime	**aSIFTime**	**[aCWmin, aCWmax]**
IEEE 802.11a	20 MHZ	9 μs	16 μs	[15, 1023]
	10 MHz	13 μs	32 μs	[15, 1023]
	5 MHz	21 μs	64 μs	[15, 1023]
IEEE 802.11b		20 μs	10 μs	[31, 1023]
IEEE 802.11g		20 μs	10 μs	[15, 1023]
		9 μs	10 μs	[15, 1023]

A station then starts its transmission if the medium stays idle for the additional duration as given by the backoff_time. Figure 3–12 depicts the contention process, including the frame transmission.

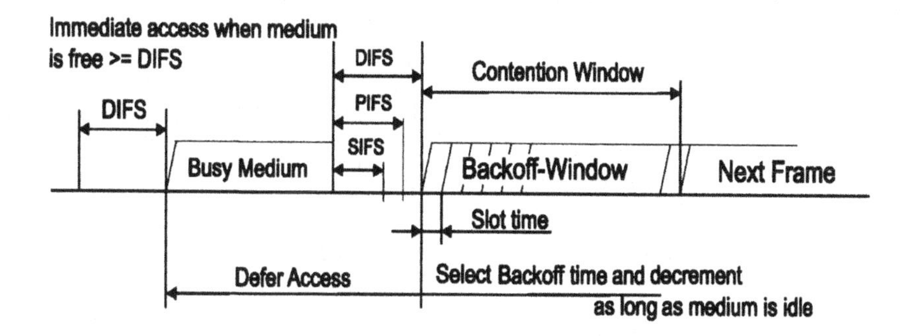

Figure 3–12: DCF Operation

After taking hold of the medium, a station will continue transmitting fragments, if necessary, of the same MSDU using short interframe spacing (SIFS) between fragments until the point when no more fragments are available or no Ack is received.

SIFS is the shortest of the interframe spacings (IFS). Using SIFS ensures that the station holds the medium for the amount of time necessary to transmit a single MSDU. Similar to aSlotTime, the value of the SIFS expressed in aSIFTime is fixed for each PHY. The aSIFTime values for the different PHY types described in this chapter are given in Table 3–2. The DIFS duration is related to the aSIFTime with the relationship:

$$\text{DIFS} = \text{aSIFTime} + 2 \times \text{aSlotTime}$$

A successful transmission is detected at the sender by the reception of an acknowledgment (Ack) control frame. A basic transmission is shown in Figure 3–13. It consists of frame transmission, a SIFS, and an Ack. The transmission of the Ack frame does not require contention for the medium. The format of the Ack frame is shown in Figure 3–14.

Absence of an Ack frame is an indication to the sender that the current transmission has failed and retransmission is needed. Crucial to the operation of the CSMA/CA is the doubling of the contention window every time transmission fails. This procedure is commonly called *exponential backoff*. The

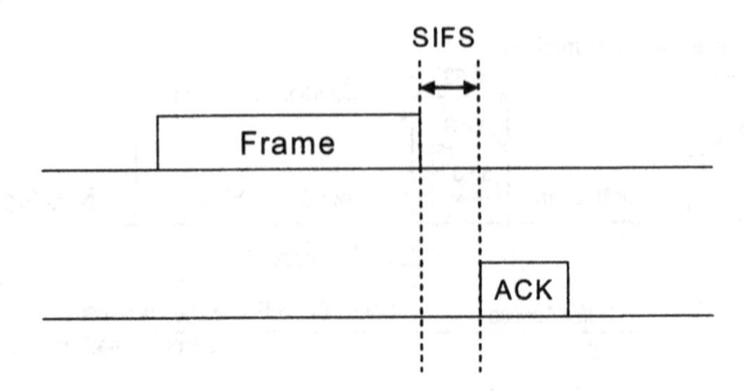

Figure 3–13: IEEE 802.11 Basic Transmission Sequence

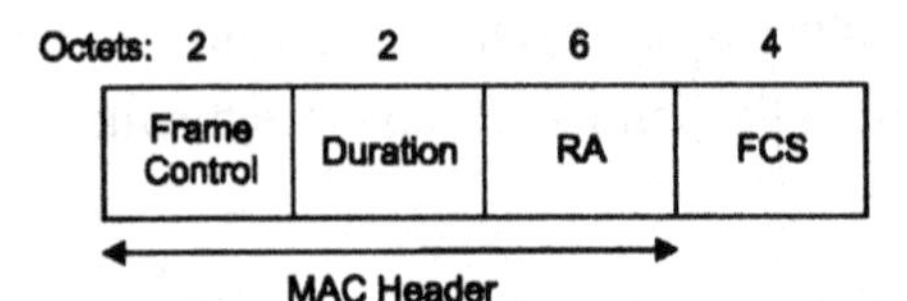

Figure 3–14: Ack Frame Format

sending station doubles the value of the contention window (cw) every time a transmission fails until it reaches aCWmax, when no further increase is allowed. The sending station has to compete for the medium with the new value for cw parameter, which effectively lowers its chances for gaining access to the medium. The values for aCWmin and aCWmax for the different PHY types are given in Table 3–2.

Carrier sensing (CS) is the operation by which the MAC layer judges the wireless medium as being busy or idle. Physical carrier sensing is carried by the PHY using the clear channel assessment (CCA) procedure. CCA detects signals on the medium, provided that the signal strength meets some minimum requirements. The result of the CCA procedure is then communicated to the MAC layer. The CCA characteristics of PHY types discussed in this chapter are given in Table 3–3.

Table 3–3: CCA Characteristics

PHY Type	PHY Parameters	CCA Characteristics
IEEE 802.11a	20 MHz Channel	> –82 dBm for 4 μs
		probability > 90%
	10 MHz Channel	> –85 dBm for 8 μs
		probability > 90%
	5 MHz Channel	> –88 dBm for 16μs
		probability > 90%
IEEE 802.11b	CCA Mode 1	Energy above threshold (> –70 dBm)
	CCA Mode 4	Detecting high rate signal during 3.65-msec interval
	CCA Mode 5d	A combination of CS and energy above threshold
IEEE 802.11g	Slot Time = 20 μs	CCA_time = 15 μsec
		Probability > 99%
	Slot Time = 9 μs	CCA_Time = 5 μsec
		Probability > 90%

IEEE Standard 802.11 provides an alternate way for carrier sensing called *virtual carrier sensing*. An IEEE 802.11 frame includes a Duration field that, if needed, can be used to indicate to other stations that the sending station intends to acquire the medium for the duration of time specified in this field. Other stations will then set their network allocation vectors (NAV) to this value and will refrain from competing for the medium during this duration.

Virtual carrier sensing is useful to deal with the hidden terminal problem. A hidden terminal is shown in Figure 3–15. In Figure 3–15 station A wants to transmit to station B. Station A is unaware that another station, station C, which is not in its transmission range, is attempting to transmit to B. If left uncontrolled, the two transmissions will collide, and transmission resources are wasted.

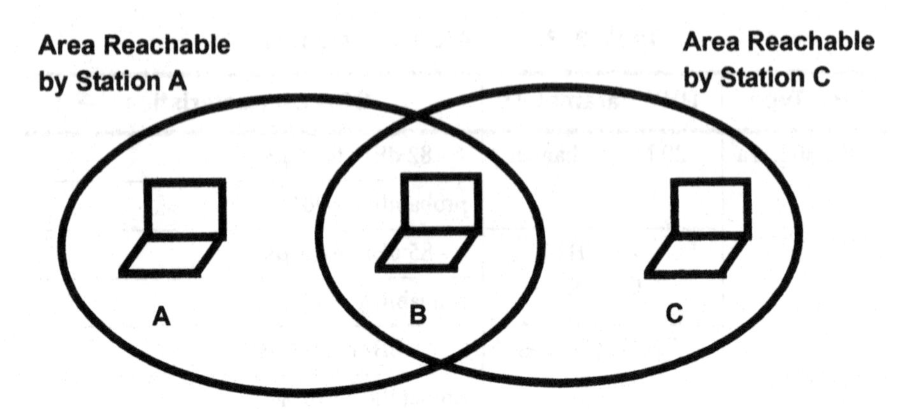

Figure 3–15: Hidden Terminal Problem

To deal with the hidden terminal problem, station A first sends a request to send (RTS) control frame to station B specifying the desired duration needed for A to complete its transmission to B. Station B will then reply to the RTS frame with a clear-to-send (CTS) control frame. Station C is able to receive the CTS frame because it is in the transmission range of station B and will refrain from sending to station B for the duration requested. Station C will achieve this objective by setting its NAV value to the desired duration. The operation of the RTS/CTS is shown in Figure 3–16.

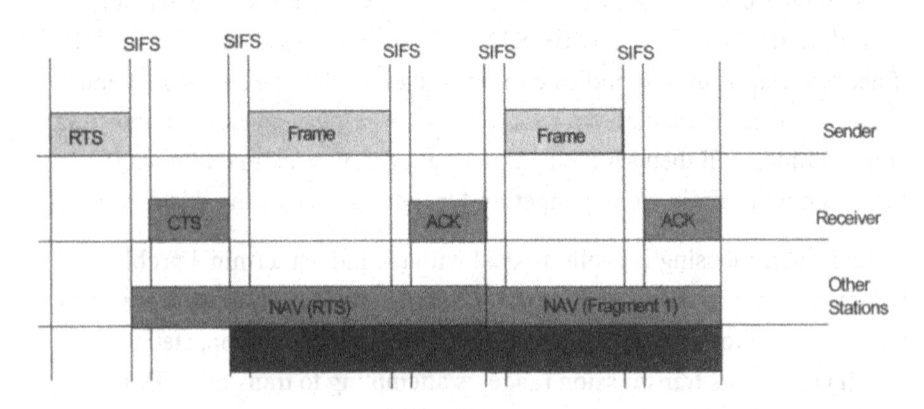

Figure 3–16: RTS/CTS Operation

Other stations monitoring the CTS/RTS messages would also set their NAV values to the value specified in the message Duration field and refrain from contending for the medium during this period. Hence the use of the RTS and CTS messages provides a mechanism to silence other stations so that transmission between station A and Station B can proceed without interruption.

The formats of the RTS and the CTS control frames are shown in Figure 3–17.

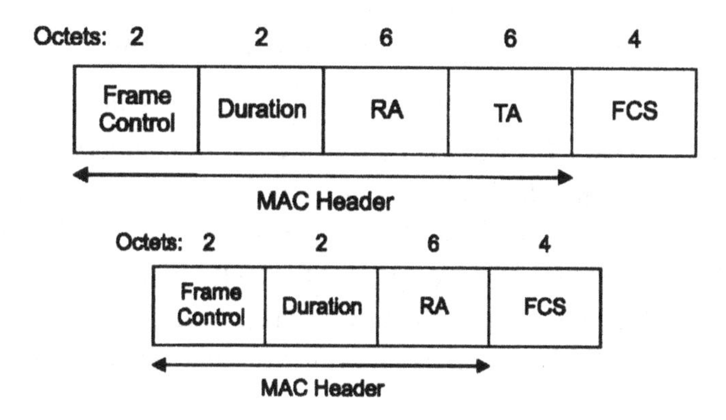

Figure 3–17: RTS and CTS Control Frames Formats

Point Coordination Function (PCF)

In addition to the basic access provided by the DCF, IEEE Standard 802.11 specifies a contention-free (CF) access mechanism based on polling. This mechanism is controlled centrally by a Point Coordination Function (PCF). The PCF usually resides at the AP. PCF is an optional IEEE 802.11 MAC feature. It provides a contention-free access to the medium based on a predetermined polling list that is stored at the point coordinator (PC).

The contention-free period (CFP) starts with the PC sensing the medium to be idle for PCF interframe spacing (PIFS) time where PIFS time is set at:

$$PIFS = aSIFSTime + aSlotTime$$

The PC then sends a Beacon management frame indicating the start of the CFP. The first Beacon and subsequent Beacon frames will include CF Parameter Set information elements that contain information relevant to the CFP operation.

The PC then starts polling stations according to its polling list, as shown in Figure 3–18. The PC can be used for polling any of these frame types: Data+CF-Poll, Data+CF-Ack+CF-Poll, CF-Poll, and CF-Ack+CF-Poll.

The CFP starts at a defined repetition rate, as indicated by the CFP repetition interval (CFPPeriod). The CFPPeriod is expressed as a number of Delivery Traffic Indication Map (DTIM) intervals. Its period is communicated to other stations in the CFP Parameter Set information element included in the Beacon frame. The length of the CFP is controlled by the PC with a maximum duration specified by the value of the CFP-MaxDuration. During the CFP, the PC transmits Beacon frames at TBTT (Target Beacon Transmission Time).

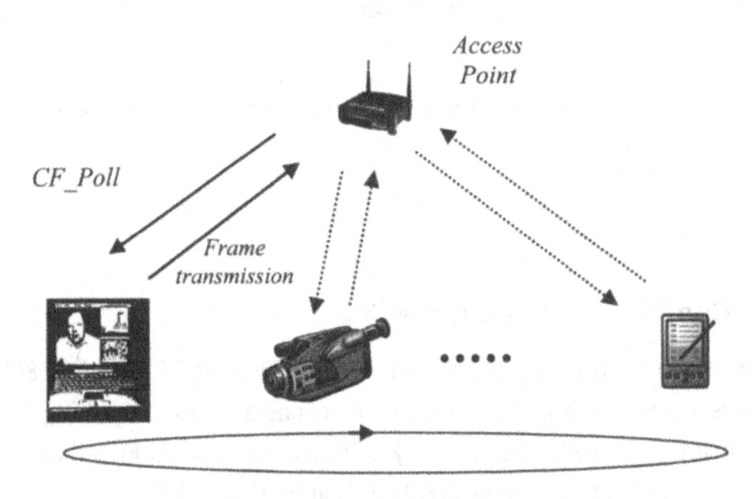

Figure 3–18: PCF Operation

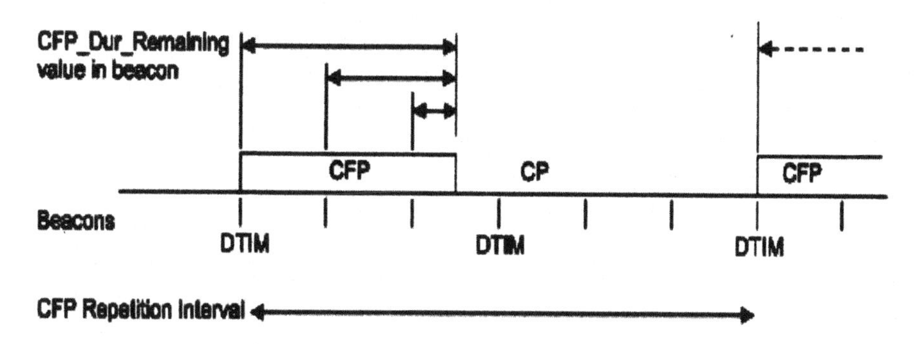

Figure 3–19: CFP and CP Alternation

The CF Parameter Set in the Beacon frames contains a nonzero value that specifies the maximum time from the most recent TBTT to the end of this CFP. This nonzero value defines the remaining time of the current CFP (CFP-DurRemaining), as shown in Figure 3–19.

The CFP alternates with the contention period (CP). The start of the CFP may be delayed from its scheduled start to allow for the completion of the current transmission, as shown in Figure 3–20.

A CF-Pollable station will respond to the PC poll by sending only one MPDU SIFS time after receiving the PC poll. The MPDU may be sent to any other station in the BSS (PC, CF-Pollable, or non-Pollable station). The station may piggyback acknowledgment for frames received from the PC.

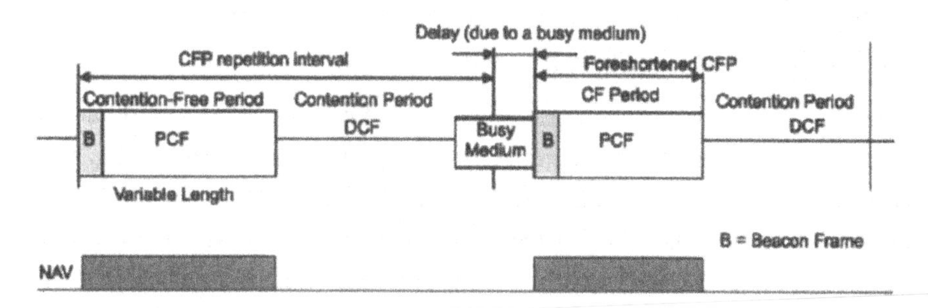

Figure 3–20: Beacon and CFP

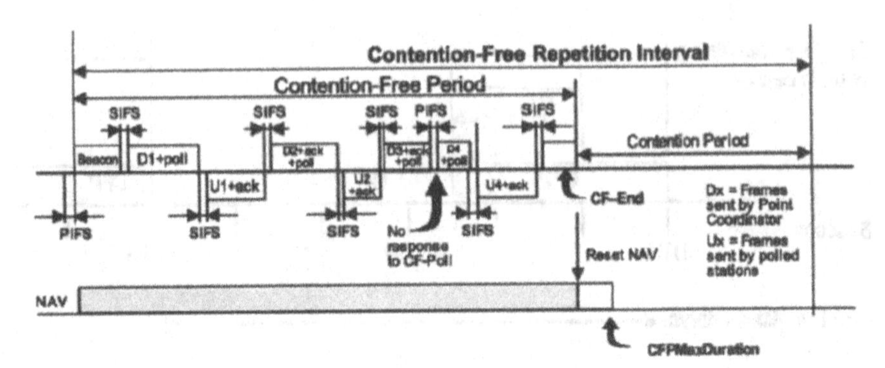

Figure 3–21: PCF Operation

A CF-Pollable station can use any of the following frame types: Data, Data+Ack, Null, and CF-Ack. A CF-Pollable station that has no frame to send when polled will transmit a Null frame. When the SIFS period elapses without the reception of an expected transmission, the PC may send its next pending transmission as soon as PIFS time elapses since last transmission, as shown in Figure 3–21. The use of SIFS and PIFS ensures that the progress of the CFP advances without interruption from non-CF-Pollable stations in the same BSS or interference from other BSS.

At the start of the CFP, all non-PC stations set their NAV to the value specified by CFP-MaxDuration. At every TBTT within this CFP, each station will update its NAV value based on the value of the CFP-DurRemaining. The end of the CFP is signaled by the PC by sending a CFP-End or CFP-Ack control frame. Stations that receive any of these frames will reset their NAV signalling the start of the contention period (CP).

Although PCF is part of the IEEE 802.11 specification for WLAN, it has not been widely implemented or deployed. WLAN deployments support the basic access mechanism as defined by DCF. Furthermore, the WLAN certification program offered by the WiFi Alliance includes only DCF-based certification. PCF is not included. WiFi certification ensures that WLAN products from different vendors can interoperate.

Power Management

Power management is one of the main features of IEEE Standard 802.11. IEEE 802.11 stations such as handheld devices or a WLAN phone are usually powered by batteries of limited operation hours. It is important that these devices have the ability to manage their power and extend operation hours.

An IEEE 802.11 station can be in one of two modes, *Awake* or *Doze*. In the Awake state the station is fully powered. In the Doze state the station is not able to transmit or receive frames and attempts to minimize its power consumption.

A station capable of power management will indicate to the AP changes in its power management mode using the Frame Control field and the Power Management bit, as shown in Figure 3–3 on page 67. An AP is not allowed to send frames to this station in an arbitrary way.

An AP includes a Traffic Indication Map (TIM) information element in its Beacon transmission. The format of the TIM IE is shown in Figure 3–22.

The Partial Virtual Bitmap is 251 bytes long (2008 bits). Bit number *N* of the Partial Virtual Bitmap field is set to 1 when the AP has traffic in its buffer ready to be transmitted for the station whose Association ID is *N*. It is set to 0 when no traffic is available.

The station using power management will periodically monitor the Beacon management frame sent by the AP. When the TIM IE of the Beacon frame indicates that traffic for this station is waiting at the AP buffer, the station sends a PS-Poll frame indicating to the AP that it is ready to receive the buffered frames. The AP then follows the normal transmission rules to forward those frames to the station.

Element ID	Length	DTIM Count	DTIM Period	Bitmap Control	Partial Virtual Bitmap

Figure 3–22: Traffic Indication Map (TIM) Information Element

Processing the TIM IE also allows stations within the BSS to be aware of the expected time of the next Delivery Traffic Indication Map (DTIM). During every DTIM Period, an AP will transmit a TIM of type DTIM within a Beacon frame rather than the ordinary TIM. The DTIM Count field indicates the number of Beacons left before the next DTIM. After a DTIM, the AP will transmit the broadcast/multicast frames using normal frame transmitting rules before transmitting any unicast frames.

IEEE 802.11 THROUGHPUT

The IEEE 802.11 constitutes a set of PHY and MAC functions as introduced in this chapter. As was discussed previously, the proper operation of the MAC and PHY functions requires the introduction of overhead fields, with obvious impact on the IEEE 802.11 throughput. Therefore, the published rates of the IEEE 802.11 devices, e.g., 11 Mbps for IEEE 802.11b or 54 Mbps for IEEE 802.11a/g cannot be achieved because of the PHY and the MAC overhead. The amount by which these rates are reduced is a function of the size of the MSDU transmitted.

The IEEE 802.11 throughput definition follows the throughput definition as given in Chapter 1 in "Performance Metrics" on page 31. It is defined as the amount of data, as measured in some units, transmitted over a given period of time. The calculation of the IEEE 802.11 throughput follows the method introduced in Jun et al. [B20] for the maximum theoretical throughput. (MTT). With this technique the system is assumed to be collision-free by having a single station always ready to transmit a new frame following the successful transmission of the preceding frame. This assumption eliminates gaps due to collisions and due to statistical variations in the traffic-arrival process.

With the previous assumption, the behavior of an IEEE 802.11 system exhibits its cyclic behavior, as shown in Figure 3–23, where each cycle is characterized by a single transmission sequence.

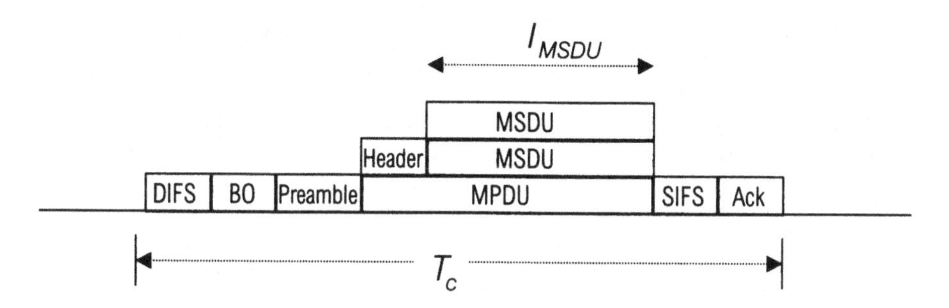

Figure 3–23: WLAN Timing Components Affecting Throughput

In Figure 3–23 a single MSDU of length l_{MSDU} bits is transmitted every T_C. Then the system throughput is given by:

$$\gamma = \frac{l_{MSDU}}{T_c} \qquad \text{EQ 3–2}$$

In addition to the time needed to transmit the MSDU, several other factors contribute to the length of T_C. Assuming the RTS/CTS mechanism is not in use, then the various components contributing to T_c duration are those shown in Table 3–4.With the assumptions made at the beginning of this section, T_C is equal to:

$$T_c = T_{DIFS} + T_{SIFS} + T_{BO} + \frac{272 + 112 + l_{MSDU}}{R} + 2 \times T_{preamble} \qquad \text{EQ 3–3}$$

The expression for T_C includes accounting for the preamble time required to transmit the MPDU and that required for Ack (not shown in Figure 3–23).

Figure 3–24 shows IEEE 802.11b throughput for different rates and MSDU sizes. As predicted, the maximum throughput achieved is less than the raw data rate, and its value is increasing monotonically with the MSDU size.

Table 3–4: Timing Components Affecting Throughput

	T_{DIFS}[a]	T_{SIFS}	T_{BO}[b]	$T_{Preamble}$	Frame Header	T_{Ack}	T_{MSDU}
IEEE 802.11a[c]	34 μs	16 μs	67.5 μs	20 μs	272[d]/R[e]	112/R	l_{MSDU}/R
IEEE 802.11b	50 μs	10 μs	310 μs	192 μs[f]	272/R	112/R	l_{MSDU}/R
IEEE 802.11g	50 μs	10 μs	150 μs	192 μs	272/R	112/R	l_{MSDU}/R

a DIFS = 2 × aSlotTime + aSIFTime

b TBO is set at ($CW_{min}/2$) × aSlotTime. This calculation assumes that the backoff time is uniformly distributed between 0 and CWmin, as specified.

c Parameter values are given for IEEE 802.11a operating in the 20 MHz.

d The IEEE 802.11 MAC frames includes 34 bytes (272 bits) of header.

e R is the rate at which the MSDU is transmitted in megabytes per second.

f In the IEEE 802.11b long PPDU format, the preamble length is 192 bits and is always transmitted at 1 Mbps. For the short PPDU format, the preamble length is 96 μs.

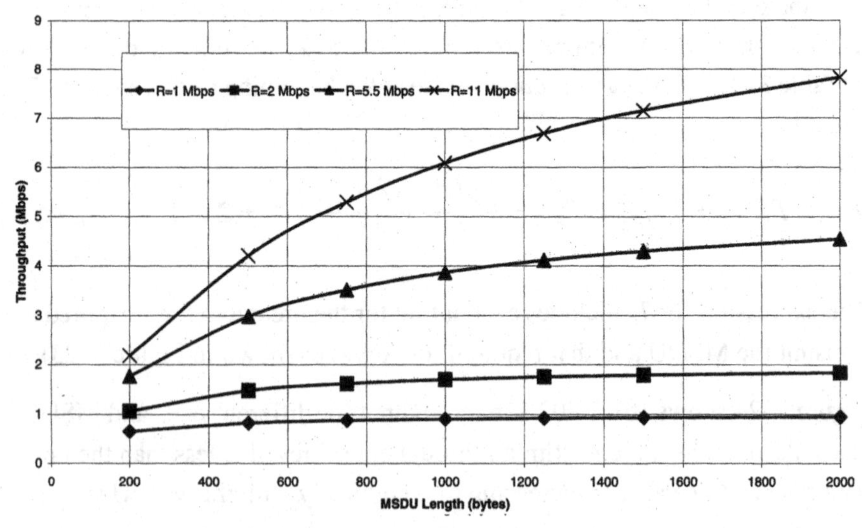

Figure 3–24: IEEE 802.11b Throughput

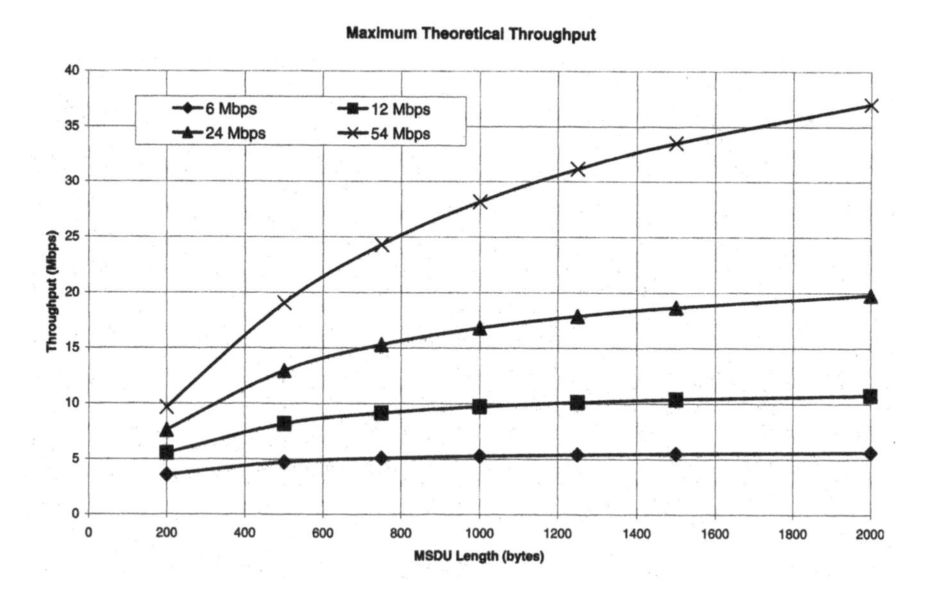

Figure 3–25: IEEE 802.11a Throughput

Figure 3–25 shows the throughput results for IEEE 802.11a. Similar to the results in Figure 3–24, the throughput results highly depends on the MSDU size. Increasing the throughput requires the use of MSDU.

Figure 3–26 shows throughput comparison when short and long preambles are used in the context of IEEE 802.11b. As expected, the use of short preamble modestly improves the throughput performance. In general, there is an improvement of approximately 8% when the short preamble is used.

Figure 3–24, Figure 3–25, and Figure 3–26 show that the straightforward way for improving WLAN throughput is to increase the length of the transmitted units relative to the preamble overhead. The High Throughput Task Group (TGn) of the IEEE 802.11 WG recommends the use of frame aggregation to achieve longer data units. Frame aggregation is possible both at the MSDU and at the MPDU levels (see "Frame Aggregation" on page 139).

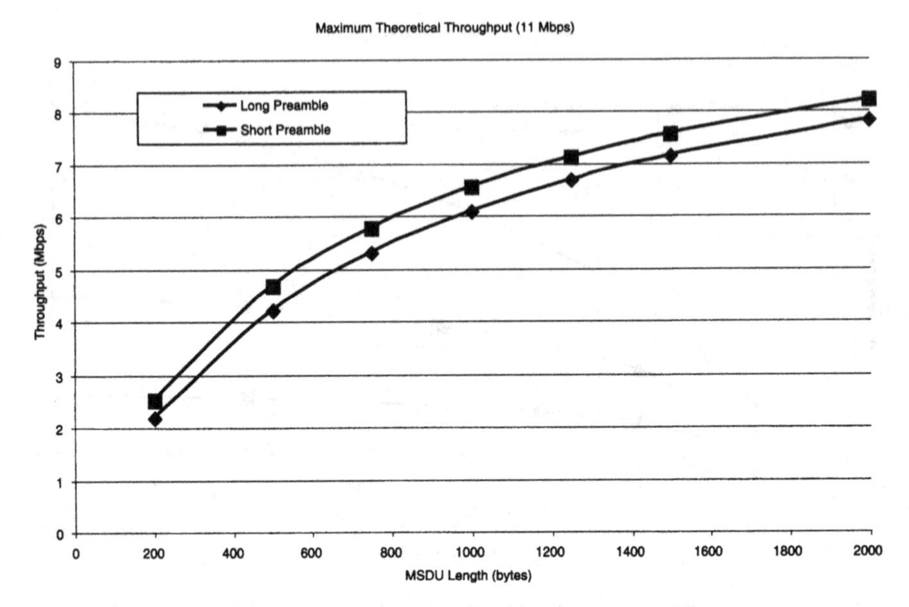

Figure 3–26: IEEE 802.11b Throughput for Long and Short Preambles

IEEE 802.11 FRAME TYPES

This section describes the coding for the different WLAN frame types. Table 3–5 includes data frame subtypes that are required for QoS operation. Most of these subtypes are related to the operation of the HCCA and do not apply to EDCA operation.

Table 3–5: IEEE 802.11 Frame Types

Type Value b3 b2	Type Description	Subtype Value b7 b6 b5 b4	Subtype Description
00	Management	0000	Association Request
00	Management	0001	Association Response
00	Management	0010	Reassociation Request
00	Management	0011	Reassociation Response

Table 3–5: IEEE 802.11 Frame Types (Continued)

Type Value b3 b2	Type Description	Subtype Value b7 b6 b5 b4	Subtype Description
00	Management	0100	Probe Request
00	Management	0101	Probe Response
00	Management	0110–0111	Reserved
00	Management	1000	Beacon
00	Management	1001	ATIM
00	Management	1010	Disassociation
00	Management	1011	Authentication
00	Management	1100	Deauthentication
00	Management	1101	Action
00	Management	1110–1111	Reserved
01	Control	0000–0111	Reserved
01	Control	1000	BlockAckReq
01	Control	1001	Block Ack
01	Control	1010	PS-Poll
01	Control	1011	RTS
01	Control	1100	CTS
01	Control	1101	Ack
01	Control	1110	CF-End
01	Control	1111	CF-End+CF-Ack
10	Data	0000	Data
10	Data	0001	Data + CF-Ack
10	Data	0010	Data + CF-Poll

Table 3–5: IEEE 802.11 Frame Types (Continued)

Type Value b3 b2	Type Description	Subtype Value b7 b6 b5 b4	Subtype Description
10	Data	0011	Data + CF-Ack+CF-Poll
10	Data	0100	Null (no data)
10	Data	0101	CF-Ack (no data)
01	Data	0110	CF-Poll (no data)
01	Data	0111	CF-Ack + CF-Poll (no data)
01	Data	1000	QoS Data
01	Data	1001	QoS Data + CF-Ack
01	Data	1010	QoS Data + CF-Poll
01	Data	1011	QoS Data+CF-Ack+CF-Poll
01	Data	1100	QoS Null (no data)
01	Data	1101	Reserved
01	Data	1110	QoS CF-Poll (no data)
01	Data	1111	QoS CF – Ack + CF-Poll (no data)
11	Reserved	0000–1111	Reserved

A Queueing Model for PCF Performance

This section demonstrates the use of the models presented in "Simple Queueing Models" on page 36 to find an approximate solution for a WLAN operating using PCF access. The model's assumptions include:

- There are N stations in the system, which are polled by the PC in a fixed order as indicated by the polling list

- Each station when polled transmits a single MPDU. The service time of MPDUs at station j is assumed to have a general distribution with mean h_j and second moment $h_j^{(2)}$.
- Arrivals of MPDU to the jth station follows a Poisson arrival process with arrival rate λ_j.
- There is U_i = SIFS time unit before the PC polls the next station. This time is called the *walking time.*

With these assumptions, the system is modeled as a multiple-queue system with ordinary cyclic service, as shown in Figure 3–27.

The analysis presented here is an approximation that provides a simple formula for computing the average waiting time at each station with a given level of the system loading. The analysis relies on the definition of the cycle time as the time elapses between two consecutive polls of station j. With this definition, station j in isolation can be modeled as in Figure 3–28. Its performance measures are those given by an $M|G|1$ queuing system with service time distribution that is equal to the distribution of the cycle time T_c.

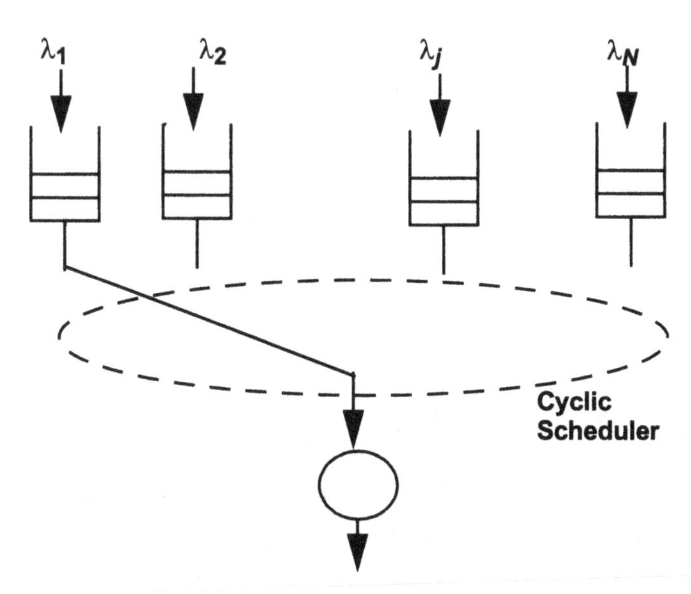

Figure 3–27: Multi-Queue Model

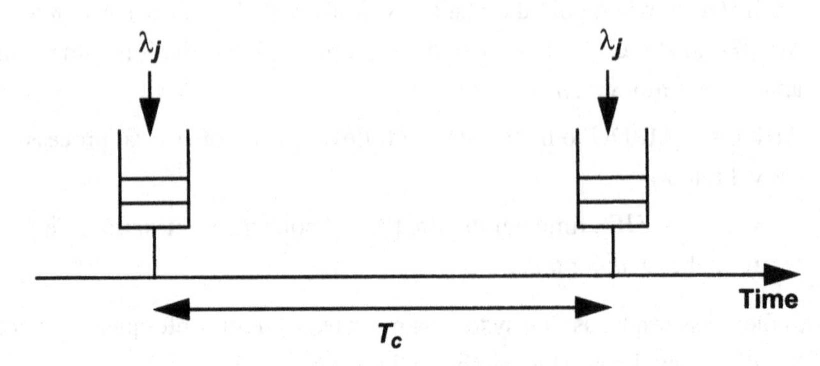

Figure 3–28: Station j in Isolation Model

Following the analysis in Kuehn [B23], the average cycle time, $c = E[T_c]$ can be easily obtained as:

$$c = \frac{c_0}{1 - \rho_0} \quad \text{EQ 3–4}$$

where:

$$c_0 = \sum_{i=1}^{N} SIFS = N \times SIFS \quad \text{EQ 3–5}$$

$$\rho_0 = \lambda_1 h_1 + \lambda_2 h_2 + \ldots + \lambda_N h_N = \sum_{i=1}^{N} \rho_j \quad \text{EQ 3–6}$$

A better approximation for the cycle time higher order statistics can be obtained by using the conditional cycle time approach presented in Kuehn [B23]. Two types of cycle times are considered. $T_c{'}_j$ is the conditional cycle time without a service from the jth queue, i.e., station j has no frames to transmit. $T_c''j$ is the conditional cycle time with a service from the jth queue, i.e., station j has a frame to transmit. The average conditional cycle times are given by:

$$c_j' = \frac{c_0}{1 - \rho_0 + \rho_j} \qquad \text{EQ 3–7}$$

$$c_j'' = \frac{c_0 + h_j}{1 - \rho_0 + \rho_j} \qquad \text{EQ 3–8}$$

The variance of the cycle time, var T_c, is given by:

$$var\ T_c = (1 - \alpha_j) \times [var\ T_{c'_j} + c'^2_j] + \alpha_j \times [var\ T_{c''_j} + c''^2_j] - c^2 \qquad \text{EQ 3–9}$$

where:

$$var\ T_{c'_j} = \sum_i varT_{U_i} + \sum_{i \neq j} (\alpha'_{ji} \times h_i^{(2)} - \alpha'^2_{ji} \times h_i^2) \qquad \text{EQ 3–10}$$

$$var\ T_{c''_j} = \sum_i varT_{U_i} + \sum_{i \neq j} (\alpha''_{ji} \times h_i^{(2)} - \alpha''^2_{ji} \times h_i^2) + var\ T_{H_j} \qquad \text{EQ 3–11}$$

$$\alpha'_{ji} = \lambda_i \times \frac{c_0}{1 - \rho_0 + \rho_j} \qquad \text{EQ 3–12}$$

$$\alpha''^2_{ji} = \lambda_i \times \frac{c_0 + h_j}{1 - \rho_0 + \rho_j} \qquad \text{EQ 3–13}$$

Knowledge of the first two moments of the cycle time allows the use of the $M|G|1$ formula for the average waiting time of a frame at the jth station. The average frame waiting time is given by:

$$E[W] = \frac{c'^{(2)}_j}{2c'_j} + \frac{\lambda_j c''^{(2)}_j}{2(1 - \lambda_j c''_j)} \quad \text{EQ 3–14}$$

Chapter 4 IEEE 802.11 QoS Features

This chapter describes in detail the QoS features added to the basic WLAN capability in an effort to support applications with diverse traffic requirements. The IEEE 802.11 WG started working on WLAN QoS extensions in 1999. It took more than five years for the task group e (TGe) to finalize its work and for the IEEE to approve what is now known as IEEE 802.11e. The length of the time needed only reflects on the difficulties related to handling QoS-related issues at the IEEE 802.11 WG. Those difficulties are by no means unique to WLAN or the IEEE 802.11. Addressing QoS and traffic management issues has always been the subject of heated debates at other standard forums, including ATM Forum, IETF, and MEF, to name a few.

Most, if not all, of the features introduced in IEEE 802.11e are centered around traffic management mechanisms that enable traffic differentiation, admission control, reservation, and other related features described in the first chapter of this book. The support for the new features necessitated the introduction of new frame formats and procedures to the WLAN standards. These additional functionality and features are described in this chapter.

In addition to the newly introduced traffic management features, the IEEE 802.11e introduced two additional features that enable the support of relatively improved throughput operations. These features are block acknowledgment (BA) and direct link setup (DLS). Both features are discussed in this chapter, and their impact on system throughput is examined.

IEEE 802.11e MAC Overview

The work of the IEEE 802.11e focused on MAC extensions only. PHY exactions were out of the scope of the TG. In reality, IEEE 802.11e MAC extensions do not depend on a particular PHY layer and should be able to make use of the service provided by any of the IEEE 802.11 PHY types.

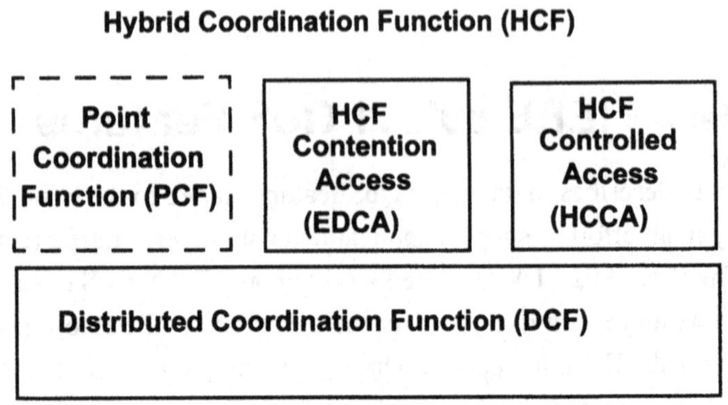

Figure 4–1: IEEE 802.11e MAC Architecture

IEEE 802.11e introduced the concept of the hybrid coordination function (HCF). The HCF now encompasses the PCF described in the previous chapter (see "Point Coordination Function (PCF)" on page 81) and two new access mechanisms, enhanced distributed channel access (EDCA) and HCF-controlled channel access (HCCA), as shown in Figure 4–1. Figure 4–1 implies that the operation of PCF, EDCA, and HCCA rely on functionality provided by the DCF, the basic IEEE 802.11 MAC.

In relationship to the QoS architectures described earlier, the EDCA access provides a mechanism for supporting traffic differentiation (or prioritization) by which real-time application traffic such as voice or video is given priority access over traffic that belongs to non-real-time applications, such as email or file transfer.

On the other hand, HCCA provides an access mechanisms that is based on a resource reservation paradigm where traffic flows are identified by a flow ID. Each flow is assigned the amount of resources requested during the admission process. Flows are rejected if resources are not available.

The introduction of the transmission opportunity (TXOP) concept ensured that a station acquiring the shared media will transmit for a finite period of time that is bounded by the length of the TXOP. A TXOP is defined as the an interval of time when a QoS station has the right to initiate frame exchange sequence onto the wireless medium. It is defined by a starting time and a maximum duration. A TXOP can be obtained by either the EDCA or the HCCA access mechanisms. Rules for acquiring the TXOP and station behavior will discussed in "EDCA Operation" on page 105 and "HCCA Operation" on page 108.

DATA FRAME FORMAT

One of the main contributions of the IEEE 802.11e is the addition of two new bytes in the data frame header, as shown in Figure 4–2. The new two bytes are called the *QoS Control* field. New data frame types are introduced to differentiate between legacy data frames and QoS data frames. The new types are QoS Data, QoS Data+CF-Ack, QoS Data+CF-Poll, QoS Data+CF-Ack+CF-Poll, and QoS Null. The new types are similar to those defined in the context of legacy station and PCF operation.

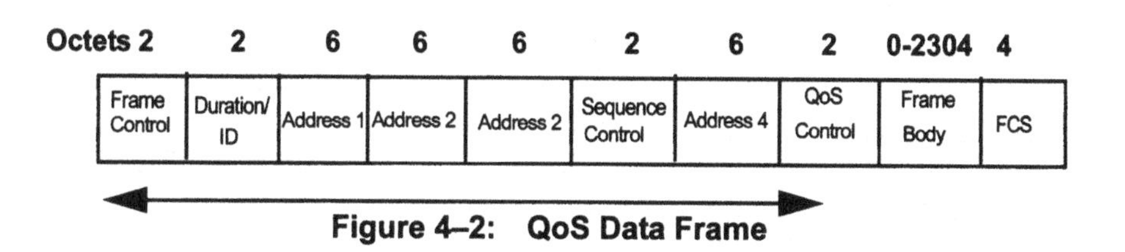

Figure 4–2: QoS Data Frame

The QoS Control field contains information necessary for the proper operation of the QoS features. The format of the QoS Control field is shown in Figure 4–3.

The first four bits of the QoS Control field are called the traffic identification (TID). When EDCA access mechanism is in use, the values 0–7 indicate the user priority (UP) of the traffic class. UP then extends the appropriate differentiation level to the associated traffic class in the same way as with wireline

Applicable Frame (sub) types	Bits 0–3	Bit 4	Bits 5–6	Bit 7	Bits 8–15
QoS (+)CF-Poll frames sent by HC	TID	EOSP	Ack Policy	Reserved	TXOP Limit
QoS Data, QoS Null, and QoS Data+CF-Ack frames sent by HC	TID	EOSP	Ack Policy	Reserved	AP PS Buffer State
QoS data frames sent by non-AP STAs	TID	0	Ack Policy	Reserved	TXOP Duration Requested
	TID	1	Ack Policy	Reserved	Queue Size

Figure 4–3: QoS Control Field

Ethernet. The UP follows those defined in IEEE 802.1D for Ethernet bridges and are summarized in Table 4–1.

Table 4–1: IEEE 802.1D Priorities

UP	Acronym	Traffic Type
1	BK	Background
2	--	not used
0	BE	Best Effort
3	EE	Excellent Effort
4	CL	Controlled Load
5	VI	Video < 100 ms latency and jitter
6	VO	Voice < 10 ms latency and jitter
7	NC	Network Control

Table 4–1 was updated with the new revision of IEEE 802.1Q [B15]. The new table is given in Table 5–1, "Priority to Traffic Types Mapping," on page 148. However, Table 4–1 is still the basis for the mapping between UP and access categories as given in Table 4–3 on page 107.

When an HCCA access mechanism is in use, the numerical values 8–15 of the four TID bits indicate the traffic stream identification (TSID). TSID in many ways is analogous to the connection identifier that is commonly used for admission control and resource reservation. A station can support up to eight different traffic streams. A TSID must be unique for each station.

The *Ack Policy* field identifies the acknowledgment policy that is followed after the delivery of an MPDU. Four Ack policies are possible: normal Ack, No Ack, No Explicit Ack, and Block Ack. The No Ack option is suitable for real-time interactive applications such as voice and video. An Ack is not required for each data frame in these types of applications, and retransmission of lost frames would only introduce undesirable delays that might render the retransmitted frame useless. The subject of Block Ack will be discussed later in this chapter (see"Block Acknowledgment (BA)" on page 129).

The *TXOP Limit* is an eight-bit field. It is included in frames sent by the AP (or the HC) to stations, and it defines in increments of 32 μs the TXOP duration granted by the AP. The range of time values is 32 μs to 8160 μs. A TXOP Limit value of 0 implies that one MPDU is to be transmitted during the TXOP.

The *Queue Size* is an eight-bit field and is included in frames transmitted from station to the AP. It describes the amount of buffered traffic for a given traffic category (TC) or traffic stream (TS). The queue size value is the total size rounded to the nearest multiple of 255 octets. The AP might use this information to determine the TXOP duration assigned to a station.

The *TXOP Duration Requested* is an eight-bit field. It is included in frames transmitted from station to the AP. It indicates the duration, in increments of 32 μs, that the sending station wants for its next TXOP for the specified TID. The AP might choose to assign a TXOP duration shorter than the one requested.

The *AP PS Buffer State* is an eight-bit field that indicate the power-saving (PS) buffer state at the AP for a station. The format of the AP PS Buffer State is shown in Figure 4–4.

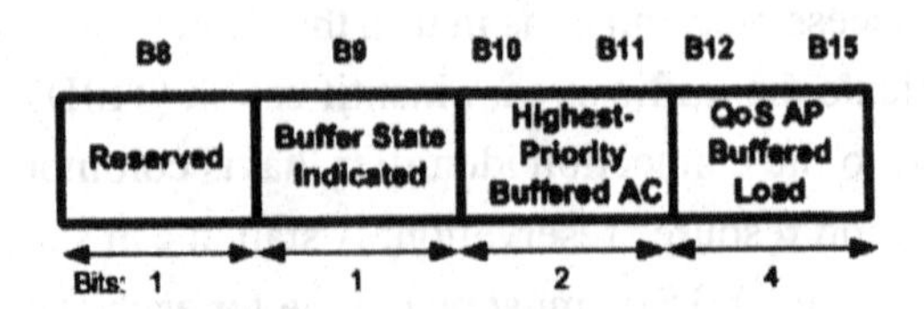

Figure 4–4: AP PS Buffer State

Buffer State Indicated subfield is one bit long and is used to indicate whether the AP PS Buffer State is specified.

Highest-Priority Buffered AC subfield is a two-bit field that indicates the access category (AC) of the highest priority traffic remaining that is buffered at the AP.

AP Buffered Load is a four-bit field that indicates the total buffer size, rounded up to the nearest 4096 octets, of all MSDU buffered at the AP.

Association to QoS BSS

A QoS-enabled station associates to a BSS in the same way as legacy stations do. Such stations send an Association Request management frame and wait for the Association Response. To facilitate the QoS operation, the Association Request management frame includes a *QoS Capability* information element (IE) sent from a station to the AP. Similarly, the Association Response includes a QoS Capability IE sent from the AP to the station. The format of the QoS Capability element when sent by a station to an AP is shown in Figure 4–5.

The first four bits of the QoS Info field indicate, per AC, whether the AC is both trigger-enabled and delivery-enabled for unscheduled power saving (U-APSD). The Q-Ack bit is set to 1 if the station is configured to receive the CF-Ack sent from the AP in response to successful transmission in QoS Data+CF-Ack destined for another station. This feature is useful to allow AP to send Ack indication for the last frames transmitted during the most recent TXOP.

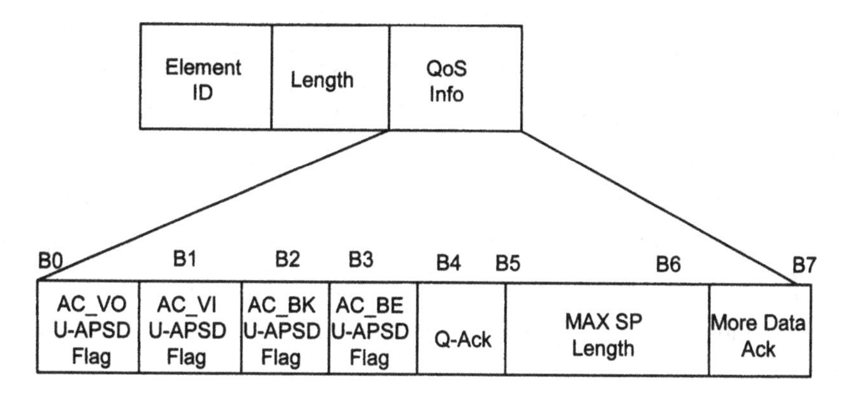

Figure 4–5: QoS Capability Information Element Sent by Station

The Max SP Length indicates the maximum number of total buffered MSDUs and MPDUs the AP can deliver to the station during a station-triggered service period.

The format of the QoS Info when sent by the AP is shown in Figure 4–6. The first four bits form the EDCA Parameter Set Update Count that is incremented every time the EDCA parameters are updated by the AP.

The AP sets the Queue Request bit to 1 if it can process a nonzero queue size in the QoS Control field in the data frames. The AP sets the TXOP Request bit to 1 if it can process a nonzero TXOP Duration Requested in the QoS Control field in the data frames.

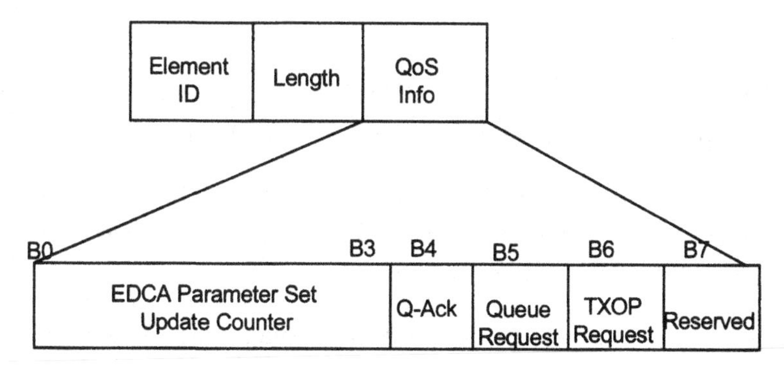

Figure 4–6: QoS Capability Information Element When Sent by AP

The Association Response management frame includes the QoS capability Information element as in Figure 4–6 as well as the *EDCA Parameter Set* IE. The EDCA Parameter Set IE is also advertised regularly by the AP in the Beacon frames.

The format of the EDCA Parameter Set IE is shown in Figure 4–7. In addition to the QoS Info field, the EDCA Parameter Set includes values for parameters to be used by stations in the BSS for EDCA access operation. One set of values is available for each AC as indicated by the two-bit AC Index (ACI) in Figure 4–7.

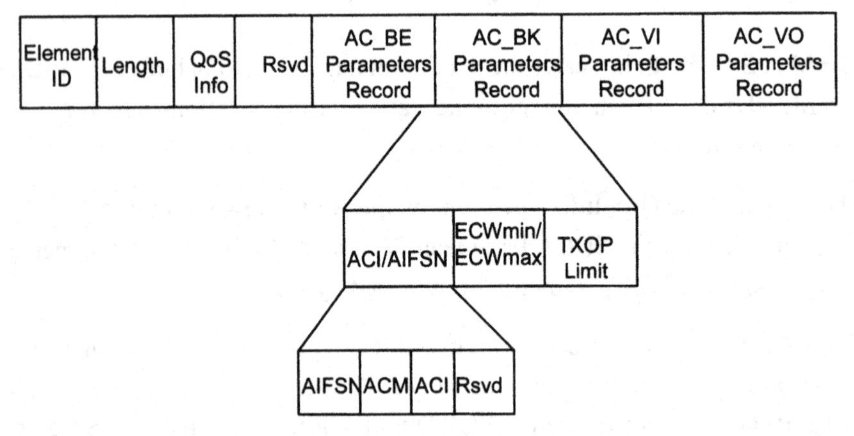

Figure 4–7: EDCA Parameter Set IE

The *Admission Control Mandatory* (ACM) bit, when set to 1, indicates that admission control is required for the AC as indicated by the ACI.

The *AIFSN* indicates the number of slots after a period of SIFS duration a station should defer before decrementing backoff or starting a transmission for an AC. The AIFSN has a minimum value of 2.

The *ECWmin* and *ECWmax* fields encode the value for the CWmin and CWmax used by an AC channel access function when competing for the wireless medium. CWmin and CWmax are computed from ECWmin and ECWmax as:

$$ECWmin = 2^{ECWmin} - 1 \qquad \text{EQ 4–1}$$

$$ECWmax = 2^{ECWmax} - 1 \qquad \text{EQ 4–2}$$

Table 4–2 shows the default parameters values for the four AC supported.

Table 4–2: EDCA Default Parameter Values

AC	CWmin	CWmax	AIFSN	TXOP Limit		
				IEEE 802.11 & IEEE 802.11a	IEEE 802.11b & IEEE 802.11g	Other PHYs
AC_BK	aCWmin	aCWmax	7	0	0	0
AC_BE	aCWmin	aCWmax	3	0	0	0
AC_VI	(aCWmin +1)/2 – 1	aCWmin	2	6.016 ms	3.008 ms	0
AC_VO	(aCWmin +1)/4 – 1	(aCWmin +1)/2 – 1	2	3.26 ms	1.504 ms	0

EDCA Operation

The EDCA defines a prioritized mechanism for wireless media access. Four access categories (AC) are defined. Each AC competes for medium with its own parameters as indicated in the EDCA Parameter Set IE and independently from other AC at the same station. Prioritized access is ensured by observing that the IEEE 802.11 MAC relies on relative timing for gaining access to the medium. For example, two stations that generate two different values for their respective backoff times would ensure that the station with the shorter time would gain access before the one with the longer time.

This same principle has been extended to allow preferred access to AC supporting real-time applications. Timely access for real-time application frames is assured by controlling the timing they wait sensing the medium. A reference implementation for the four EDCA AC is shown in Figure 4–8.
Figure 4–8 shows that a QoS-enabled station runs four separate channel access functions, one for each AC. The operation of each channel access function is similar to that of the DCF for legacy stations.

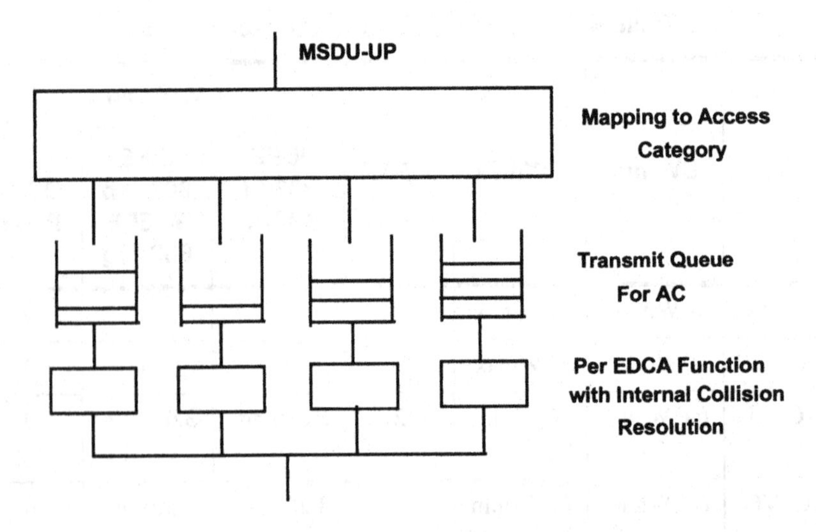

Figure 4–8: Reference Implementation Model

The four EDCA access categories are labeled AC_VO (AC Voice), AC_VI (AC Video), AC_BE (AC Best Effort), and AC_BK (AC Bulk). One AC is differentiated from another based on the values of parameters it uses to access the medium. The access category of an incoming frame is determined by the frame UP as indicated in the TID field of the QoS Control field. A one-to-one mapping is defined between the UP and the AC, as shown in Table 4–3.

Table 4–3: UP to AC Mapping

UP	802.1D Designation	AC	Designation
1	BK	AC_BK	Background
2	—	AC_BK	background
0	BE	AC_BE	Best Effort
3	EE	AC_BE	Best Effort
4	CL	AC_VI	Video
5	VI	AC_VI	Video
6	VO	AC_VO	Voice
7	NC	AC_VO	Voice

As shown in Figure 4–9, a station listens to the medium for a deterministic time called Arbitration Inter-Frame Spacing (AIFS). The AIFS is related to the AIFSN in the EDCA Parameter Set IE. This relationship is given by:

$$AIFS[AC] = AIFSN[AC] \times aSlotTime + aSIFSTime \qquad \text{EQ 4–3}$$

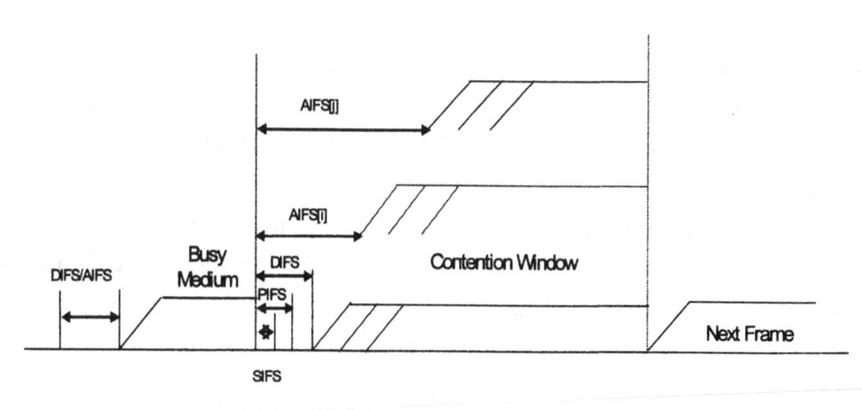

Figure 4–9: EDCA Channel Access Functions

In the absence of a collision, the result of the contention process is that a station would acquire the medium for a particular AC. At this time the station is said to acquire an EDCA TXOP. The duration of the TXOP is limited by the TXOP Limit as advertised by the AP. The duration of the TXOP may be sufficient to send more than one MSDU. If this is the case, then the station is allowed to use its TXOP to send frames only from the same AC for which the TXOP is acquired. A TXOP Limit of 0 implies that the station is allowed to transmit one transmission sequence including SIFS and Ack. A station should time its transmission so that it never exceeds the TXOP duration allocated.

Failed transmission as indicated by the absence of an Ack will trigger a backoff. EDCA backoff follows the same backoff procedure as in legacy IEEE 802.11. In backoff, the contention window is doubled if necessary every time there is a collision and is reset back to its CWmin when transmission is successful.

Backoff procedure is also invoked when internal collision occurs. Internal collision occurs when two or more EDCAFs in the same station are granted TXOP at the same time. The station in this case allocates the TXOP to the highest priority AC. Other colliding ACs within the same station will invoke backoff and double the size of its contention window if necessary.

HCCA Operation

Similar to the PCF, HCCA provides controlled, contention-free access to the shared medium. HCCA differs from PC in two aspects. HCCA, unlike PC, can be invoked at any time even during the contention period. Also, associated to HCCA is an admission control procedure that ensures availability of resources for HCCA flows. Furthermore, HCCA and the controlled access phase always start at predefined instants as determined by the scheduler. On the other hand, PC start can be delayed waiting for the current transmission to be completed, as shown in Figure 3–20 on page 83. All these factors contribute to the predictability of the service offered to real-time flows.

HCCA operation is centrally controlled by the HC. Similar to the PC, the HC is co-located with the AP. The HC architecture includes a scheduler that determines for each of the HCCA flows the amount of service it receives and when it receives this service. These calculations will require the ability of the HC to initiate a controlled access phase (CAP) whenever appropriate. The ability of the HC to interrupt an ongoing CP and start a CAP is shown in Figure 4–10.

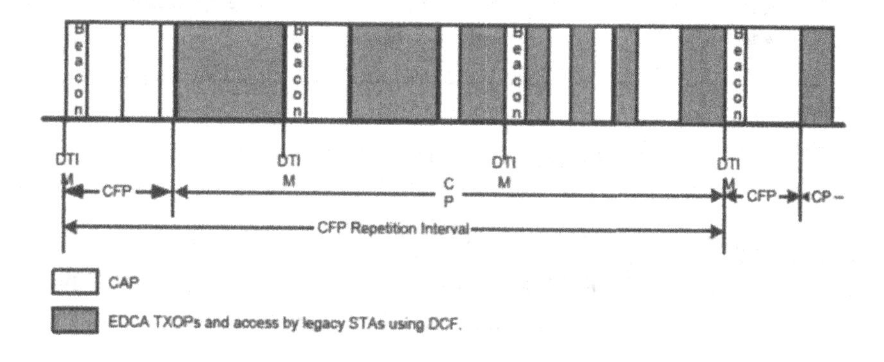

Figure 4–10: Controlled Channel Access

As in the PC case, the first permitted frame at the start of the CFP is a Beacon frame. The CAP ends when the TXOP ends and the HC does not reclaim the medium after PIFS duration.

The ability of HC to start CFP at any time differs from the PC where CFP and CP alternate almost periodically. Periodic switching between CFP and CP has the effect of introducing additional delay that might degrade the performance of real-time traffic. For example, a voice frame arriving at t_1 has to wait until the start of the next CFP under the operation of the PC. With HC operation, the transmission of the incoming voice frame can start shortly after its arrival, as shown in Figure 4–11.

When HC gains control of the medium, it starts polling stations according to the requirements of the traffic streams they support. It uses QoS + CF-Poll

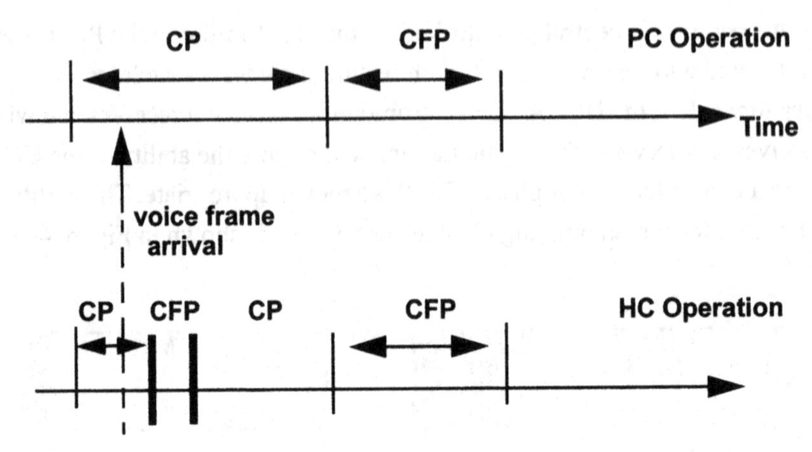

Figure 4–11: HC and PC

frames for this purpose. Unlike the PC operation, the HC does not keep a polling list to follow. HC polling is directly coupled to traffic stream requirements and allocation as determined by the admission procedure.

During a TXOP, the HC or a station sets its NAV to prevent its transmission during the duration of the TXOP. The TXOP duration is included in the QoS Control field of the +CF-Poll frame. Stations set their NAV according to the contents of the Duration/ID field received frame.

If there is no more data to transmit or the HC has finished polling all eligible stations, the HC can end the CFP by sending a QoS CF-Poll frame with the RA matching its own MAC address with Duration/ID field set to 0. The reception of this frame by the different stations within the BSS will cause them to reset their NAV to 0, hence allowing them to start contending for the media if needed.

Admission Control

The IEEE 802.11e TG included admission control as one of the key components of a QoS architecture. The TG has introduced the mechanisms and the supporting protocol elements necessary for enabling the implementation of admission control functionality in a WLAN environment supporting QoS.

As discussed in "Admission Control" on page 11 in Chapter 1, admission control requires the exchange of information between the end user (a station in WLAN) and the admission control module. The required exchange of information is achieved in IEEE 802.11e by introducing Action management frames of the appropriate types. The general format of Action frames is shown in Figure 4–12.

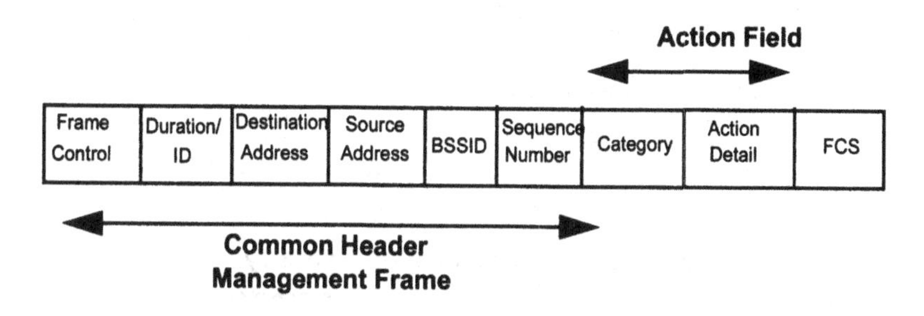

Figure 4–12: Action Frame Format

Like any management frame, the action frame starts with the management frame common header. The Action field is in the management frame body and consists of:

- The one-byte Category field that defines the category of the action specified
- A variable size action detail part that specifies the action and the supporting elements.

The action categories specified so far are shown in Table 4–4.

The first byte that follows the action category identifies the action type in the context of the specified category. Up to 256 different actions can be specified in each category. The Action frame types relevant to QoS are shown in Table 4–5.

Table 4–4: Action Categories

Code	Category
0	Spectrum Management
1	Quality of Service (QoS)
2	Direct Link Setup (DLS)
3	Block acknowledgment (BA)
4–126	Reserved
127	Vendor Specific
128–255	Error

Table 4–5: QoS-Related Action Frame Types

Action Field Value	Action
0	ADDTS Request
1	ADDTS Response
2	DELTS
3	Schedule
4–255	Reserved

Figure 4–13 shows the initiation of a generic admission control procedure. The procedure starts with the transmission of ADDTS (ADD Traffic Stream) Request Action frame.

The ADDTS Request includes a TSPEC IE describing traffic requirements and performance of the request. The ADDTS Request-related elements are shown in Figure 4–14. The ADDTS Request frame may optionally include a TCLAS IE. The TCLAS IE includes a set of parameters necessary to identify the incoming MSDU from a higher layer perspective. For example, Classifier Type 1 includes elements related to IPv4 parameters such as source and destination IP addresses, DSCP, and port number.

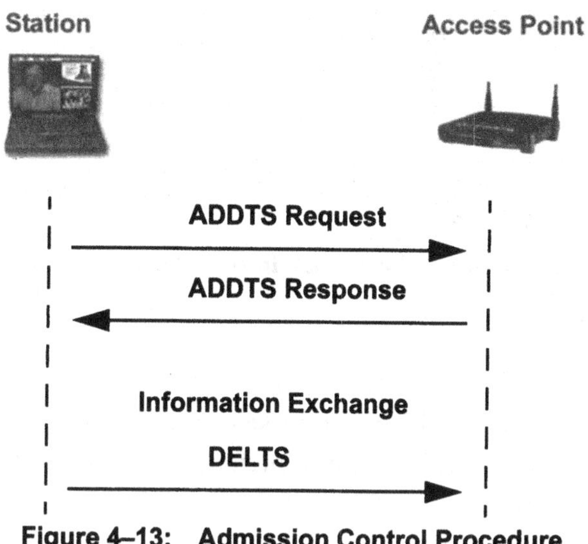

Figure 4–13: Admission Control Procedure

Category	Action	Dialog Token	TSPEC	TCLAS	TCLAS Process

Figure 4–14: ADDTS Request Frame Format

The AP responds to ADDTS Request with an ADDTS Response frame that includes the result of the admission process and a TSPEC IE that includes those parameters that are currently being supported. Those parameters may be different from the one transmitted with the ADDTS Request. The format of the ADDTS Response Action frame is shown in Figure 4–15.

Category	Action	Dialog Token	Status Code	TS Delay	TSPEC	TCLAS	TCLAS Processing	Schedule

Figure 4–15: ADDTS Response Frame Format

Element ID	Length	TS Info	Nominal MSDU Size	Maximum MSDU Size	Minimum Service Interval	Maximum Service Interval	Inactivity Interval	Suspension Interval

Service Start Time	Minimum Data Rate	Peak Data Rate	Burst Size	Delay Bound	Minimum PHY Rate	Surplus Bandwidth Allowance	Medium Time

Figure 4–16: The TSPEC Information Element

TSPEC IE is crucial for the admission process. The format of the TSPEC IE is shown in Figure 4–16. The TSPEC IE includes parameters relevant to the expected traffic characteristics, required performance, and parameters relevant to the service process.

The structure of the TS Info field is three bytes long and is shown in Figure 4–17. The following definitions describe the different subfields of the TS Info.

Traffic Type is a single-bit field that indicates the traffic as being either periodic or aperiodic. When the Traffic Type bit is set to logic 1, it indicates that traffic is expected to be periodic with constant interarrival times between frames (MSDU). Unlike constant bit rate traffic (CBR), the MSDU can be different lengths. When Traffic Type is set to logic 0, it indicates aperiodic traffic.

TSID is a four-bit field that always describes the traffic stream ID. As described before in the context of QoS Control field, the most significant bit (MSB) of the TSID subfield is always set to 1. The TSID is unique within a station. A single station can identify up to eight traffic streams.

Traffic Type	TSID	Direction	Access Policy	Aggregation	APSD	User Priority	TS Info Ack Policy	Schedule	Rsvd

Figure 4–17: The TS Info Field

Direction is a two-bit field that describes the direction of the traffic carried by the TS as shown in Table 4–6.

Table 4–6: Direction Subfield

Bit 1	Bit 2	Usage
0	0	MSDUs are sent from station to HC
1	0	MSDUs are sent from HC to station
0	1	Direct link MSDUs are sent from one station to another station
1	1	Bidirectional link (equivalent to a downlink request and an uplink request. Each direction has the same parameters.

Access Policy is a two-bit subfield that specifies the access that would be used to access the medium according to Table 4–7.

Table 4–7: Access Policy Subfield

Bit 1	Bit 2	Usage
0	0	Reserved
1	0	Contention-based channel access (EDCA)
0	1	Controlled channel access (HCCA)
1	1	HCCA, EDCA mixed mode (HEMM)

Aggregation subfield is a single bit in length. It is used only when the access policy is HCCA or EDCA and the schedule subfield is set to1. A station sets the subfield to 1 to indicate that an aggregate schedule is required. When it is set to 1 by the AP, it indicates that an aggregate schedule is provided to the station. It is set to 0 otherwise.

APSD is a single-bit subfield, and it is set to 1 to indicate that automatic power saving (PS) delivery is to be used for traffic associated with the TSPEC and is set to 0 otherwise.

User Priority (UP) is a three-bit subfield that indicates the relative priority used for the transmission of MSDU belonging to the TS if prioritization is required.

TSInfo Ack Policy is a two-bit subfield that indicates whether MAC acknowledgments are required for MPDU belonging to the TS and the desired form of acknowledgments. Table 4–8 shows the allowed Ack policies.

Table 4–8: TSInfo Ack Policy

Bit 1	Bit 2	Usage
0	0	Normal IEEE 802.11 Acks
1	0	No Ack
0	1	Reserved
1	1	Block Ack

Schedule is a single-bit subfield. It is used only when the access policy is set to EDCA to indicate the type of the APSD, Scheduled or Unscheduled, as shown in Table 4–9. When Schedule and APSD are set to 1, the Aggregation field will also be set to 1 by the AP, indicating that an aggregate schedule is being provided to the station.

Table 4–9: Setting of Schedule Subfield

APSD	Schedule	Usage
0	0	No schedule
1	0	Unscheduled APSD

Table 4–9: Setting of Schedule Subfield

APSD	Schedule	Usage
0	1	Reserved
1	1	Scheduled APSD

Nominal MSDU Size is a two-byte long field. The MSDU can be of fixed or variable nominal size as indicated by the fixed bit. The remaining 15 bits contain an unsigned number that specifies the nominal size.

Maximum MSDU Size is a two-byte long field. It includes an unsigned integer that specifies the maximum size in bytes of the MSDUs belonging to the TS.

Minimum Service Interval is a four-byte field that contains an unsigned integer specifying the minimum interval, in microseconds, between the start of two successive service periods.

Maximum Service Interval is a four-byte field that contains an unsigned integer specifying the maximum interval, in microseconds, between the start of two successive service periods.

Inactivity Interval is a four-byte field. It contains an unsigned integer that specifies the minimum amount of time, in microseconds, that may elapse without arrival or transfer of an MPDU belonging to the TS before it is deleted by the HC.

Suspension Interval is a four-byte field. It contains an unsigned integer that specifies the minimum time, in microseconds, that may elapse without arrival or transfer of an MSDU belonging to the TS before the generation of successive +CF-Poll is stopped for this TS. the value of the suspension interval is always less that or equal to that of inactivity interval.

Service Start Time is a four-byte field containing an unsigned integer. It specifies the time, in microseconds, when the first scheduled service period starts.

Minimum Data Rate is a four-byte field. It contains an unsigned integer that specifies the minimum acceptable data rate at the MAC service access point (MAC_SAP) for the TS. The minimum rate is expressed in bits per second and does not include the MAC and the PHY overheads incurred in transferring the MSDU.

Mean Data Rate is a four-byte field. It contains an unsigned integer. It specifies the average data rate specified at the MAC_SAP for the TS. The mean rate is expressed in bits per second and does not include the MAC and the PHY overheads incurred in transferring the MSDU.

Peak Data Rate is a four-byte field. It contains an unsigned integer. It specifies the maximum allowable data rate specified for the TS. The peak rate is expressed in bits per second.

Burst Size is a four-byte field that contains an unsigned integer. It specifies the maximum burst, in bytes, belonging to the TS that arrives at the MAC_SAP at the peak data rate.

Delay Bound field is a four-byte field that contains an unsigned integer. It specifies the maximum amount of time, in microseconds, allowed to transport and MSDU belonging to the TS. The time is measured from the instant an MSDU arrives at the local MAC sublayer from the local MAC_SAP until the instant when the frame successfully reaches its destination. It include the relevant Ack frame transmission time if needed.

Minimum PHY Rate is a four-byte field that contains an unsigned number. It specifies the desired minimum PHY rate to use for the TS. The rate is specified in bits per second.

Surplus Bandwidth Allowance specifies the allocation of time and bandwidth over the stated application rates required to transport an MSDU belonging to the TS in this TSPEC. It takes into account the additional resources needed for retransmission of lost or corrupted frames.

Medium Time is a 16-bit field that contains an unsigned integer. It contains the amount of time admitted to access the medium in increments of 32 ms/s. It is used for prioritized access only and indicates the amount of time a station is allowed to access the medium for a particular access category.

In the previous definitions, the peak data rate is defined so that in any interval of length Δt, where Δt is greater than a single time unit (TU) of length 1024 μs, the amount of information arriving does not exceed peak rate multiplied by Δt.

The mean data rate and the burst size are defined using the token bucket model discussed in "The Metering Function" on page 5. Because the burst size parameter represents the maximum burst size allowed, the token bucket depth L is set equal to:

$$B = \frac{L}{1 - \frac{p}{R}} \qquad \text{EQ 4–4}$$

where B is the maximum burst size, p is the source peak data rate, and R is the maximum PHY rate.

EDCA Admission Control

Admission control in the EDCA mode of operation is allowed only for access categories for which the HC declares that admission control is mandatory. Setting the ACM bit to logical 1 in the EDCA Parameter Set IE makes this declaration. In the case where the HC declares that admission control is not required for any of the access categories, then no EDCA admission control is needed.

EDCA admission control is performed per station and per access category. It is not performed per traffic stream because the concept of traffic stream and its identifier are not supported in EDCA operation.

EDCA admission control starts by the station issuing ADDTS Request Action frame to the HC. The station specifies the appropriate TSPEC parameters including the UP associated with the AC that requires admission control. It is recommended that the station specify the Nominal MSDU Size, Mean Data

Rate, Minimum PHY Rate, Inactivity Interval, and Surplus Bandwidth Allowance parameters. Other TSPEC parameters are not set and are ignored.

The HC receives the TSPEC and decides whether to accept or to deny the incoming request. If the HC decides to accept the incoming request it must specify the Medium Time parameter in the TSPEC IE sent back to the station using ADDTS Response Action frame.

The procedure at the station implements what is usually referred to as a use-it-or-lose-it procedure. The station keeps two counts, *admitted_time* count and *used_time* count, which initially are set to logical 0.

The admitted_time count is updated with value of the medium time allowed by the HC, in increments of 32 μs, in dot11EDCAveragingPeriod. The used_time is the amount of time used, in increments of 32 μs, by the station in dot11EDCAveragingPeriod for accessing the medium.

On the receipt of the TSPEC IE with Medium Time specified, the station updates the admitted_time count as follows:

> admitted_time = admitted_time + dot11EDCAveragingPeriod × (Medium Time)

At dot11EDCAveragingPeriod second Intervals:

> used_time = max ((used_time – admitted_time), 0)

After each successful or unsuccessful MPDU transmission or retransmission attempt:

> used_time = used_time + MPDUExchangeTime

The station should refrain from accessing the medium with the specified AC when used_time reaches or exceeds the admitted_time. The station can continue to access the medium at an AC that has a lower priority than the one specified in the admission.

The effect of this procedure is that the time is divided into fixed-length intervals of length dot11EDCAvergingPeriod, as shown in Figure 4–18. When the

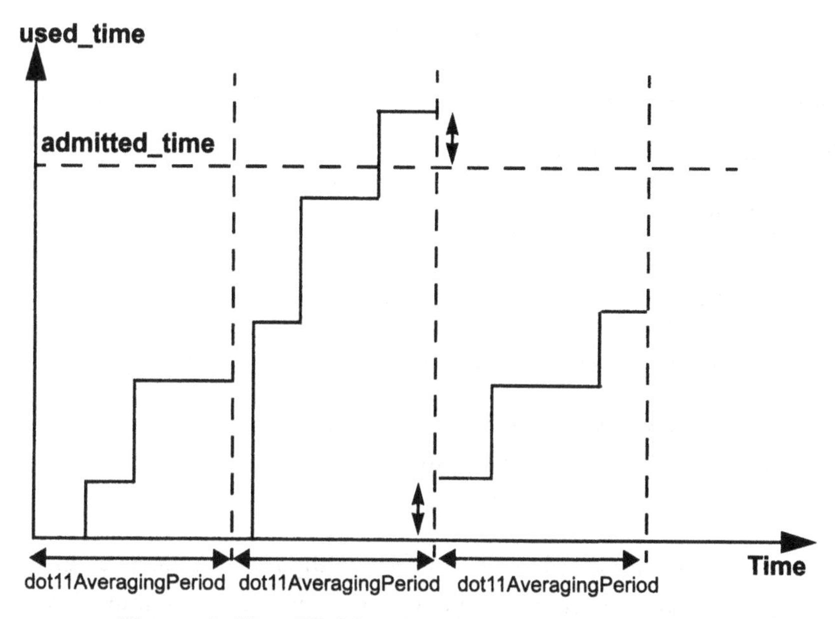

Figure 4–18: EDCA Admission Control Behavior

used_time is less than the admitted_time, the used_time is reset to 0 at the next interval. When the used_time exceeds the admitted_time, the used_time in the next interval is set to the difference between the used_time and the admitted time, as is shown in Figure 4–18. This procedure has the effect that any left over used_time from one interval is simply discarded at the start of the next interval, hence the name "use it or lose it."

The computation of the Medium Time by the HC based on parameters included in the TSPE IE is not a trivial problem. A suggested mechanism for setting the Medium Time is included in IEEE 802.11e [B17]. It sets the Medium Time equal to the average number of MSDU expected (at the Nominal Size) during an MPDUExchangeTime scaled by the Surplus Bandwidth Allowance according to the relationship:

$$T_{medium} = SBW \times \frac{\lambda}{8 \times L} \times T_{Exchange} \qquad \textbf{EQ 4–5}$$

where:

T_{medium} = Medium Time

SBW = Surplus Bandwidth Allowance

λ = Mean Data Rate in bits per second

L = Nominal Data Size in bytes

$T_{Exchange}$ = MPDU ExchangeTime defined as the time needed for a single data exchange at the Nominal Data Size and the minimum PHY rate.

The previous expression for the Medium Time seems to be conservative, especially for the cases where MPDUExchangeTime is significantly shorter than the dott11EDCAveragePeriod.

Equation 4–5 calculates the medium time. It totally relies on the average rate and average size with no consideration for the variations in the frame length distribution and its arrival process. The impact is that during periods of temporarily high rates, a backlog might form and frames may be dropped because the assignment of the medium time relies only on average parameters.

Information about the peak data rate is needed in order to account for the variability in the data arrival process. If peak rate information is available, you can replace the parameter λ in Equation 4–5 with $\lambda^{\sim}$ given by:

$$\tilde{\lambda} = \frac{2 \times \lambda}{1 + \theta} \qquad \textbf{EQ 4–6}$$

where θ is the activity factor of the traffic source and is defined as the probability that the traffic source is active. Note that $\lambda^{\sim}$ is always greater or equal to the λ. This relationship tends to increase the value of the computed medium time to better handle traffic variations.

The peak data rate is not one of the parameters specified for the EDCA admission control. However, a network administrator can configure those parameters that allow better management of the network.

HCCA Admission Control

HCCA admission control is done on a per traffic stream (TS) basis. The HC is responsible for scheduling channel access to TS, based on information included in the TSPEC IE. The HCCA admission control procedure starts with the station sending an ADDTS Request Action frame including the TSPEC IE to the HC. The station indicates that the access policy is HCCA, and it sets the TSID for the TS for which reservation is requested. The TSID is unique within the station.

The stations sets TSPEC parameters to enable the HC to make admission decision. A minimum set of TSPEC parameters include Mean Data Rate, Nominal MSDU Size, Minimum PHY Rate, Surplus Bandwidth Allowance, and at least one of Maximum Service Interval and Delay Bound. The TSPEC parameters setting should be driven directly from application parameters and requirements. For example, in the case of constant bit rate (CBR) voice at 64 Kbps, the mean data rate is set at application rate of 64 Kbps. In this case the minimum data rate and the peak data rate are equal to the mean data rate because of the continuous nature of the voice traffic.

The HC sends an ADDTS Response Action frame back to the requesting station. The ADDTS Response includes a Schedule IE. The Schedule IE includes information on how often the station will be polled for this TS and for how long the HCCA TXOP will last. The format of the Schedule IE is shown in Figure 4–19.

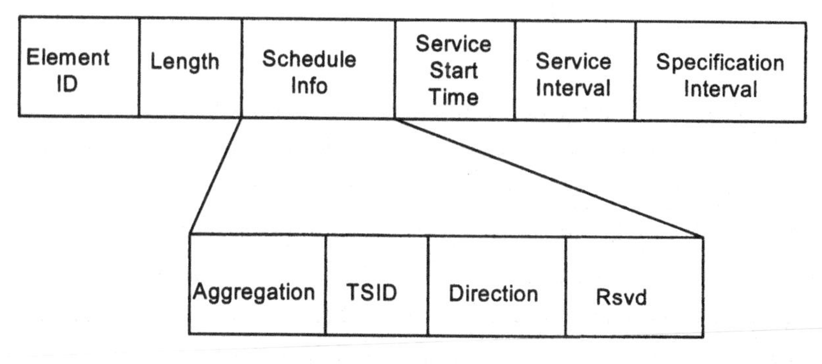

Figure 4–19: Schedule IE Format

Guidelines were given in IEEE 802.11e [B17] on how to set the scheduler service interval. It can be set to the Minimum Service Interval parameter value if it is one of the parameters specified in the TSPEC IE. Alternatively, it can be set equal to the nominal MSDU size divided by the mean data rate, which makes the assumption that MSDU of average size are arriving at a rate equal to the mean data rate.

The guidelines for setting the TXOP duration suggest that the TXOP duration is set equal to the time needed to transmit an average number of MSDU that arrive during one service interval. In particular:

$$TXOP = max\left(\frac{\left\lceil \frac{SI \times r}{L} \right\rceil \times L}{R} + O, \frac{M}{R} + O\right) \qquad \text{EQ 4–7}$$

where:

SI = Service Interval

r = Mean Data Rate

L = Nominal MSDU Size

R = Minimum PHY Rate

M = Maximum MSDU Size

O = Overhead

The first term on the left side of Equation 4–7 represents the average number of MSDU arrivals during a service interval at the average rate and the nominal data size. As was mentioned before, assigning resources based on average statistics does not take into account the variabilities in the arrival process, especially for cases where application traffic is bursty in nature. Using average statistics could lead to underallocation of resources in this case, and hence performance values might be worse than anticipated.

An alternate way for computing the service schedule by the HC is to use Equation 1–4 on page 14. The equation takes into account the burstiness of the traffic by using a two-state Markovian model for the arrival process. The use of the equation requires knowledge of the desired frame loss rate. The loss rate is not one of the parameters signaled in the TSPEC IE. It might be provided as a vendor-specific field, or it can be configured by the WLAN operator. The mapping between equation parameters and the parameters signaled in the TSPEC IE is given by:

β^{-1} = Maximum Burst Size / Peak Data Rate

γ = Peak Data Rate

α^{-1} = (Peak Data Rate – Nominal Data Rate)/(β × Nominal Data Rate)

B = Buffer size at the AP

The outcome of the EQ will be in the form of an equivalent bandwidth or effective bit rate (EBR) in units per second, e.g., bits per second. With the information about the service interval and the PHY rate, it is possible to compute the TXOP duration as:

$$TXOP = \frac{EBR \times SI}{R} \qquad \text{EQ 4–8}$$

Direct Link Setup

Direct link setup (DLS) is a mechanism that allows two WLAN stations in an infrastructure to communicate together in a direct way without the need to send their frames first to the AP to be relayed to the destination. DLS is introduced as part of IEEE Standard 802.11e. DLS has the potential to improve the BSS throughput by a factor of 2. It can be used in almost all scenarios where AP intervention is not needed, such as VoIP and video transmission, where the two-end stations are in the active mode and do not need the AP to store their frames.

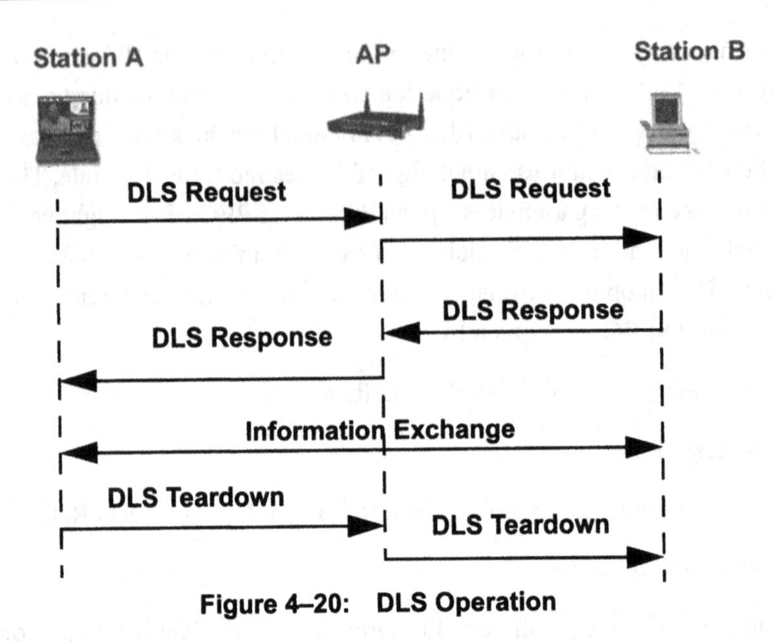

Figure 4–20: DLS Operation

The operation of the DLS is shown in Figure 4–20. The DLS operation starts when station A attempts to set up a direct link to station B. Station A sends a DLS Request Action frame to the AP. The AP forwards the DLS Request to the destination station. Station B responds with a DLS Response Action frame. The DLS Response frame includes a status code that indicates the result of the DLS Request. The AP relays the DLS Response frame back to the source.

If the DLS setup is successful, information exchange between station A and station B can now proceed directly without the need to direct every frame transmission to the AP first. At the end of the information exchange the DL can be torn down by either station by sending DLS Teardown Action frame.

Similar to the WLAN QoS, the operation of the DLS is facilitated by using of Action management frames with a Category field equal to DLS. The DLS Action types supported are given in Table 4–10.

Table 4–10: DLS Action Frame Types

Action Field Value	Action
0	DLS Request
1	DLS Response
2	DLS Teardown
3–255	Reserved

The format of the DLS Request frame is shown in Figure 4–21. The originating station sets the source MAC Address field to its MAC address. It sets the destination MAC address to the MAC address of the station with which it wants to establish a DL. The DLS Request also includes a DLS Timeout Value field. The DL is terminated if it remains inactive for a period of time equal to the value of this field. A value of 0 indicates that the DL never terminates because of inactivity.

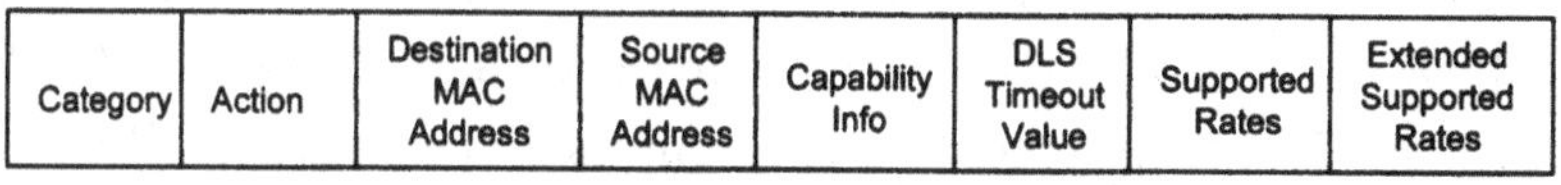

Category	Action	Destination MAC Address	Source MAC Address	Capability Info	DLS Timeout Value	Supported Rates	Extended Supported Rates

Figure 4–21: DLS Request Frame Structure

The AP relays the DLS Request frame to the destination station, as shown in Figure 4–20. If the destination station is willing to accept the request, it responds back to the AP with a DLS Response frame. The DLS Response frame structure is shown in Figure 4–22. The status code indicates the outcome of the setup procedure. If the status code indicates success, then the target station includes its Capability Information in the DLS Response frame.

After the direct link between the two stations is established, transmission can proceed without first going through the AP, as shown in Figure 4–20. A station can use any of the IEEE 802.11 access mechanisms for the direct link. If

Category	Action	Status Code	Destination MAC Address	Source MAC Address	Capability Info	DLS Timeout Value	Supported Rates	Extended Supported Rates

Figure 4–22: DLS Response Frame Structure

AP will then poll the station at the scheduled service instance. The station is then free to send its frames during the specified TXOP directly to the target station as specified by the DLS setup.

In those cases where EDCA is used for medium access, the two stations involved in the direct link transmission must use the EDCA parameters advertised by the AP.

Teardown of the direct link can be initiated either by a station or by the AP by sending a DLS Teardown Action frame. Receiving the DLS Teardown by the AP or a station is an indication to close the DLS and remove all the associated states. The DLS Teardown frame structure is shown in Figure 4–23.

The DLS mechanism does not grant more QoS privileges other than those defined by EDCA and HCCA operation. In setting a direct link, the AP does not reject the DLS Request for lack of resources. In fact, the DLS Request Action frame does not carry any indication of the resources needed for the DLS Request. The AP can reject a DLS Request only for reasons related to validity of the DLS parameters and whether direct link operation is supported by the target station. On the other hand, the AP may reject the QoS features associated with a DL. This association is performed independently from the DLS procedure.

Category	Action	Destination MAC Address	Source MAC Address	Reason Code

Figure 4–23: DLS Teardown Frame Structure

Block Acknowledgment (BA)

Block acknowledgment (BA) is a mechanism for allowing the acknowledgment of multiple MPDUs as one block instead of acknowledging each MPDU separately. BA can improve WLAN throughput by removing the need to individually acknowledge each frame. Removing the need for individual Ack allows a station to wait SIFS units between successive frame transmissions. Two types of BA are defined, immediate BA and delayed BA.

Two BA-capable stations can start a BA peer relationship by exchanging the ADDBA Request and ADDBA Response Action management frames, as shown in Figure 4–24.

The BA-related actions are shown in Table 4–11.

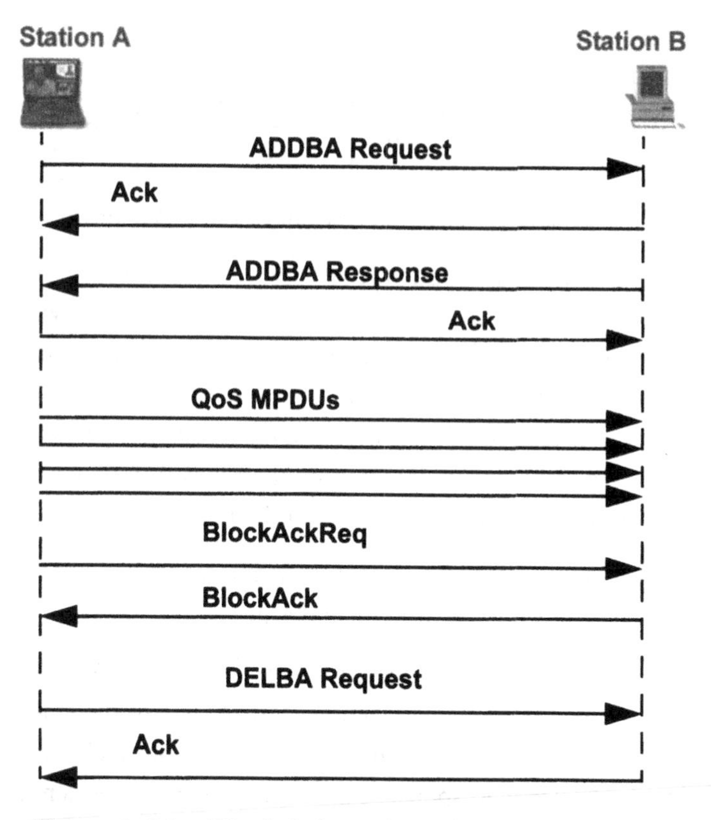

Figure 4–24: Block Acknowledgment Procedure

Table 4–11: BA Action Frame Types

Action Field Value	Action
0	ADDBA Request
1	ADDBA Response
2	DELBA
3–255	Reserved

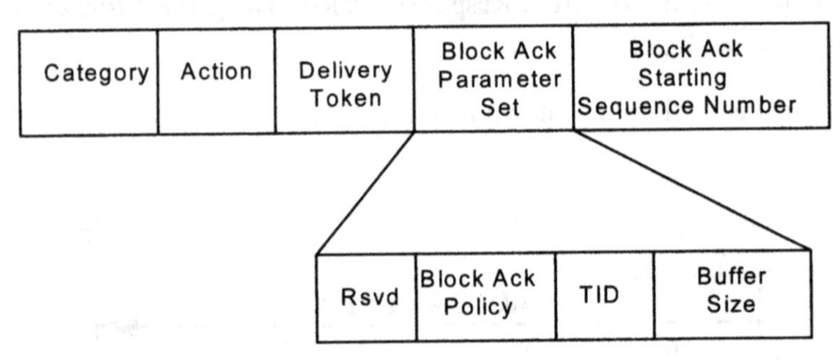

Figure 4–25: ADDBA Request Frame Structure

The ADDBA Request frame structure is shown in Figure 4–25. It includes the Block Ack Parameter Set field. This field includes the TID for which the BA setup is applied. It also includes the Buffer Size subfield, which indicates the number of buffers in increments of 2304 available for this BA. In the ADDBA Request frame, the Buffer Size subfield is only advisory and can be rejected or modified by the recipient of the request. The Block Ack Policy bit indicates which BA policy is applicable: immediate BA or delayed BA.

The Block Ack Starting Sequence Control field is 12 bits long. It indicates the first MPDU sequence number for which BA is applied.

The receiver of the ADDBA Request responds with an ADDBA Response Action management frame. The structure of the ADDBA Response is shown in Figure 4–26. The value of the Buffer Size in the Block Ack Parameter Set included in the ADDBA Response is the actual value that the sender has to use and is the one that can be supported by the receiver.

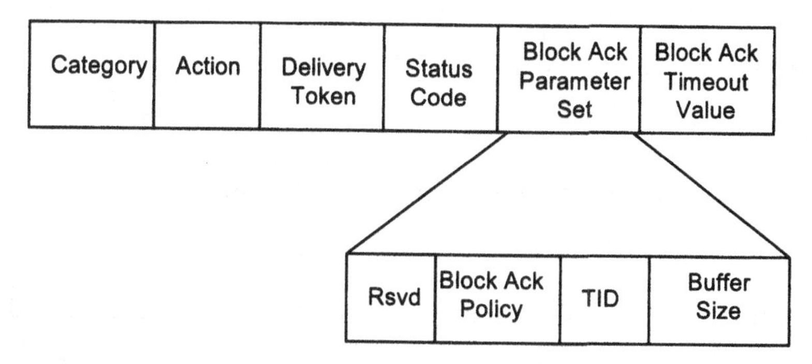

Figure 4–26: ADDBA Response Frame Structure

After a BA relationship is established between two stations, the originator will send a block of MPDUs up to the Buffer Size. The sending of the MPDUs can be extended over a number of TXOP. The TXOP may be obtained by EDCA or HCCA access mechanisms as usual. MPDUs that are sent during one TXOP are separated by SIFS.

The sender request the acknowledgment of the block of MPDUs that have already sent by sending BlockAckReq control frame to the receiver. The format of the BlockAckReq control frame is shown in Figure 4–27. RA and TA are the MAC addresses of the recipient and the transmitting stations. The frame also includes the Block Ack Starting Sequence Number field, which indicates the starting sequence number for which the BlockAck Request applies. The BAR Control field includes the TID of the flow for which the block acknowledgment applies.

The receiver of the BlockAckReq frame responds with a BlockAck control frame. The format of the BlockAck control frame is shown in Figure 4–28. It includes the same Block Ack Starting Sequence Number received in the BlockAckReq frame.

Frame Control	Duration ID	RA	TA	BAR Control	Block Ack Starting Sequence Number	FCS

Figure 4–27: BlockAckReq Control Frame Format

Frame Control	Duration ID	RA	TA	BAR Control	Block Ack Starting Sequence Number	Block Ack Bitmap	FCS

Figure 4–28: BlockAck Frame Format

The acknowledgment of the various MPDUs in the block is included in the Block Ack Bitmap field. The length of the Block Ack Bitmap field is 128 bytes and is used to indicate the receiving status of up to 64 MSDUs. When the nth bit of the Block Ack Bitmap field is set to 1, it indicates that the MPDU with the sequence number equal to the Block Ack Starting Sequence Number + n is received without errors. Otherwise, it indicates that the reception of the MPDU has failed.

In addition to the BA operation described previously, IEEE 802.11e has also introduced the No Ack policy for frames where reliability at the link layer is not an issue or for applications where the delay introduced by the need for retransmission may deem the frame out of date. This might be the case in most real-time applications such as VoIP and IPTV.

Throughput Improvement Due to Block Acknowledgment

As was mentioned before, the main contribution of the block acknowledgment mechanism is the opportunity for improving the system throughput by shortening the time needed to transmit consecutive MPDU. In evaluating the contribution of the block acknowledgment mechanism for improving throughput, the same assumptions are used for the evaluation of IEEE 802.11 throughput presented in Chapter 3, "IEEE 802.11 Throughput" on page 86.

Figure 4–29 shows the basic transmission sequence used to evaluate the system throughput. Figure 4–29 shows the case for two frames that are transmitted SIFS units apart. The transmission of the two frames is followed by a BA acknowledging the two frames. The BA is transmitted following the reception of BlockAckRequest, as shown in Figure 4–24 on page 129. The situation can

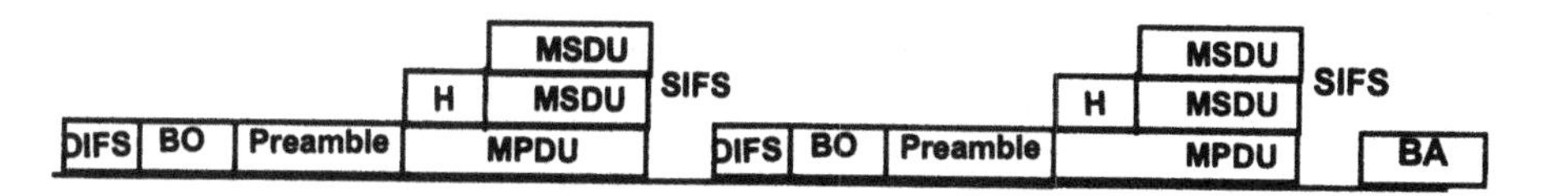

Figure 4–29: Block Ack Throughput

be extended to the case where n frames are transmitted where n is less or equal to 64. In general, when n frames are transmitted, the time needed to complete the task is given by:

$$T_c = T_{DIFS} + T_{BO} + n \times \left(T_{preamble} + \frac{272 + l_{MSDU}}{R}\right) + \frac{l_{BAR} + l_{BA}}{R} + 2T_{preamble} + (n-1) \times T_{SIFS} \quad \text{EQ 4–9}$$

The expression for T_C includes the times needed to transmit the BlockAckReq and the BlockAck control frames and the associated preamble.

The throughput is given by:

$$\gamma = \frac{n \times l_{MSDU}}{T_c} \quad \text{EQ 4–10}$$

Figure 4–30 shows the throughput improvement when Block Ack is applied as a function of the number of the MSDU transmitted before requesting the Block Ack control frame. Obviously, when $n = 1$, the throughput of a BA system is lower than the normal Ack. However, the BA procedure is not recommended for the $n = 1$ case. The throughput then increases as n increases and improves compared to the normal Ack system. Figure 4–30 was obtained with the assumption that the TXOP is long enough to allow the transmission of the desired number of MPDUs. This length seems to be the main factor for improving the throughput because a station does not have to contend for the medium using DIFS and backoff.

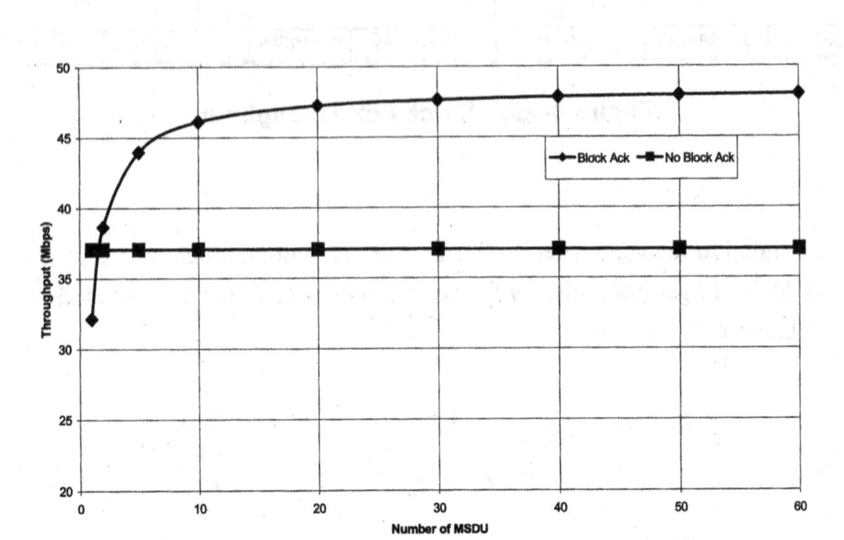

Figure 4–30: Block Ack Throughput Improvement

Power Management Using APSD

IEEE 802.11e added Automatic Power Save Delivery as a way for managing the delivery of downlink frames to power-saving (PS) stations. A station indicates its power management mode in the power management field of the Frame Control field. The AP will then start buffering frames for stations in the PS mode. The AP announces stations for which frames are buffered periodically in the Beacon frame.

The APSD defines two delivery mechanisms, unscheduled APSD (U-APSD) and scheduled APSD (S-APSD). S-APSD starts at fixed intervals of time as specified in the Service Interval field. A station indicates its intent to use S-APSD by sending to the AP an ADDTS Request with the APSD subfield of the TS Info field of the TSPEC IE set to 1. When EDCA mechanism is used for channel access, the station will also set the Schedule subfield of the TS

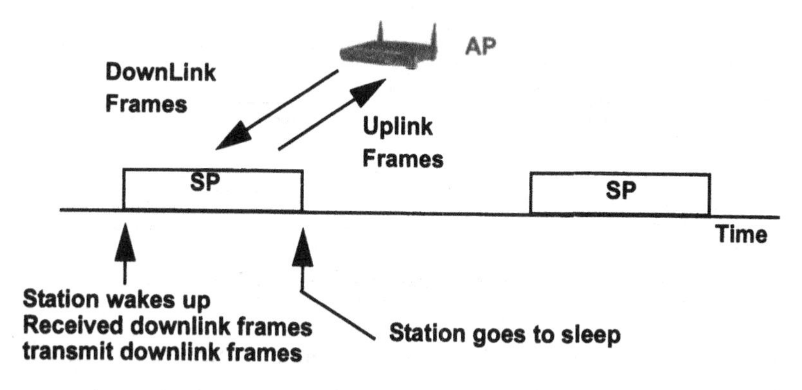

Figure 4–31: Scheduled APSD

Info field of the TSPEC IE to 1. If the AP supports S-APSD, it responds back with an ADDTS Response frame including the Schedule IE, indicating the service interval and the service start time.

A PS station wakes up at the start of the service period (SP) as scheduled. During the SP the station receives downlink frames, if any are stored at the AP, and transmits uplink frames. The station will then go to sleep at the end of the SP. The process is repeated at the start of every SP, as shown in Figure 4–31.

U-APSD is used when EDCA is the channel access mechanism in use. In order to configure an AP to deliver frames during an unscheduled SP using U-APSD, it needs to configure one or more of its AC to be triggered-enabled and delivery-enabled. Setting the corresponding U-APSD Flag subfield in the QoS capability IE will set the AC as both delivery-enabled and trigger-enabled. Otherwise, the AC is enabled neither for trigger nor delivery.

Alternatively, a station can use the ADDTS and TSPEC IE to set an AC as either delivery-enabled or trigger-enabled. Setting the APSD and Schedule subfields in the TS Info field will configure the AC.

With U-APSD a station does not need to wake up at the start of predefined SP to send and receive frames. Instead, an unscheduled SP may start at any time.

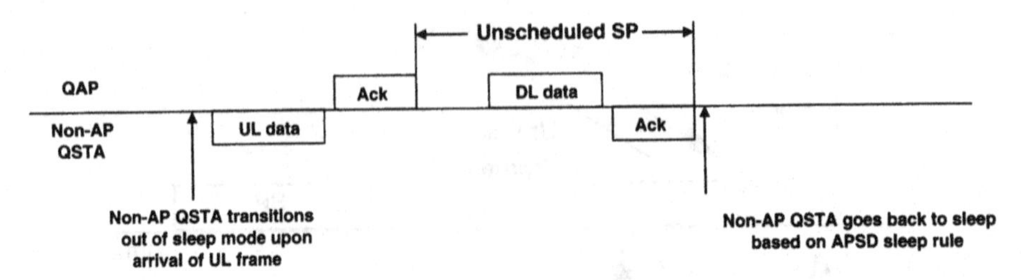

Figure 4–32: Unscheduled APSD

The start of the unscheduled SP begins when the AP receives a trigger frame from the station. This frame is a QoS Data or QoS Null frame associated with an AC that the station has declared to be trigger-enabled.

Figure 4–32 shows the operation of the U-APSD. The station wakes up with the arrival of an uplink frame. The arrival of the uplink frame triggers the start of unscheduled SP. The trigger frame sent to the AP indicates that the station is out of the sleep mode. The AP can then send the station the downlink frames stored in its buffer. The unscheduled SP ends after the AP attempts to transmit at least one MSDU to the station, but no more than the number indicated in the Max SP Length field if the field has a nonzero value.

WiFi Multimedia (WMM) Specification

WLAN products from different vendors are required to interoperate smoothly. The WiFi Alliance is concerned with specifying interoperability testing. The WiFi Alliance usually designs its interoperability test scenarios and specification based on IEEE 802.11 WG drafts and approved standards. Therefore, the WiFi Alliance does not advocate any major departures from IEEE 802.11 WG recommendations.

In specifying the interoperability testing for QoS based on IEEE 802.11e, the WiFi Alliance followed the normal procedure. However, because of the long time taken by the IEEE 802.11e TG to publish its final recommendation and market pressure to act quickly on vendor interoperability of QoS features, inevitably some discrepancies exist between the WiFi Alliance QoS specification, WMM [B32], and the approved IEEE 802.11e specification. The WiFi

Alliance QoS certification program is based on WMM specification. Therefore, WMM is widely implemented by many WLAN vendors.

WMM is based on the EDCA functionality of the IEEE 802.11e specification. Similar to the IEEE 802.11e, the WMM supports four access categories. WMM access categories have the same names and the same UP mapping as in IEEE 802.11e.

The WMM specification uses the same frame format, as shown in Figure 4–2. However, the QoS Control field recommended by WMM is different from the one proposed by the IEEE 802.11e in Figure 4–3. The WMM QoS Control field is shown in Figure 4–33.

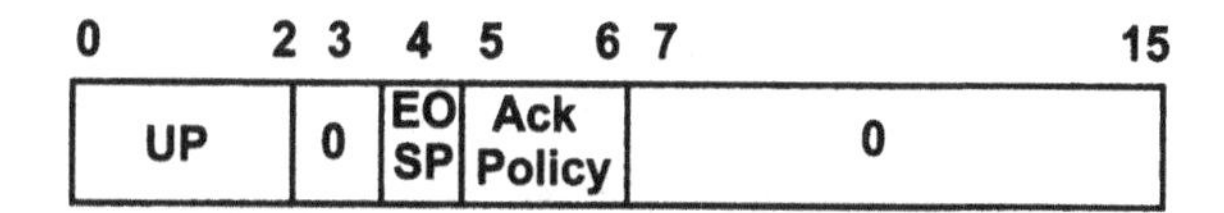

Figure 4–33: WMM QoS Control Field

Differences between the two formats of the QoS Control field seem to be motivated by the fact that the WMM specification provides no support for the HCCA access function. Hence all fields needed for HCCA are left unspecified or with a different scope. For example, IEEE 802.11e format supports a four-bit TID with the ability to include the TSID in the data frame header, but the WMM format supports three-bit UP field, which is sufficient for the function supported.

Furthermore, the WMM specification ignores all data frames subtypes such as QoS Data+CF-Poll because they have no use in EDCA channel access operation. The WMM specification uses only the additional data subtypes QoS Data and QoS Null.

In addition to the EDCA basic operation, the WMM specification supports the related admission control. WMM EDCA admission control is slightly different from that of IEEE 802.11e. WMM Admission control procedure follows the same rules specified in IEEE 802.11e by using ADDTS Request and

ADDTS Response action frames with the TSPEC IE, which has the same format in the two specifications. However, WMM admission control imposes a limit of one TS per AC at any point in time. A station can aggregate its traffic and make it fit into a single TS. The time reference for the Medium Time allocated for an AC in WMM specification is 1-second intervals. The reference time in IEEE 802.11e is dot11EDCAAveragingPeriod, which can be configured to values other than one second.

The format the QoS Information field needed to manage the QoS operation in a BSS is also influenced by the lack of WMM support for the HCCA channel access. Figure 4–34 shows the format of the QoS Info field when sent from the AP and from the station.

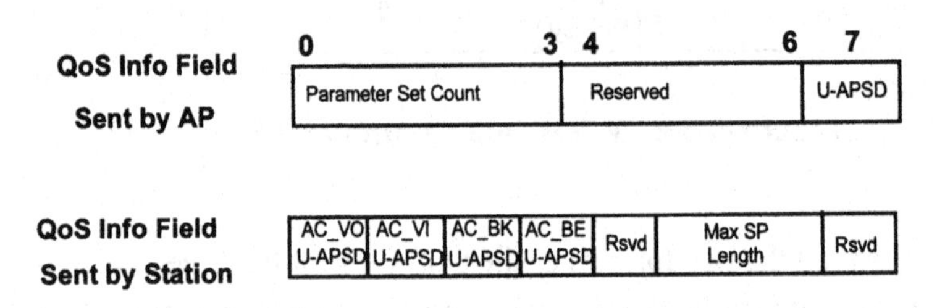

Figure 4–34: WMM QoS Control Field

Block acknowledgment and direct link setup are not supported by the WMM specification. Two Ack policies are supported, Acknowledged and Do Not Acknowledge. WMM also supports the unscheduled APSD. No support for scheduled APSD is included.

An IEEE 802.11e device is incompatible with a WMM device. Only one of these capabilities is required to be advertised during association. As of the time this book was written, an effort was underway at the IEEE 802.11 WG to harmonize WMM and IEEE 802.11e specifications. It is still early to predict the outcome of this level.

IEEE 802.11N (HIGH THROUGHPUT) MAC FEATURES

The High Throughput (HT) Task Group (TGn) was formed in 2003 with the mandate to specify MAC and PHY enhancements to support a throughput of at least 100 Mbps at the MAC data service access point (SAP). The final specification has not been approved yet. However, the approach taken by the TG for increasing the throughput has matured to the point where it can be described here.

The TGn has added a number of PHY and MAC options. If all options are implemented, a throughput of approximately 600 Mbps can be achieved. This section describes the main MAC enhancements to support high throughput.

MAC enhancements introduced by the TGn have not added fundamentally to the QoS and traffic management features introduced by TGe. TGn recognizes that a high throughput station is also a QoS station in the sense that it supports access mechanisms described in IEEE 802.11e and uses the tools introduced there in terms of traffic and stream identifiers, user priority, mapping to access categories, and admission control. At the same time, TGn has modified the use of some of the IEEE 802.11e features, especially the use of the block acknowledgment.

Frame Aggregation

Recognizing that long frames are necessary for achieving high throughput, TGn allows frame aggregation at the MSDU and at the MPDU levels. These frame aggregations are referred to as A-MSDU and A-MPDU, respectively.

The A-MSDU format is shown in Figure 4–35. Figure 4–35 also shows the general high throughput data frame format, including the HT Control field. The Frame Body can be up to 7955 bytes long.

The existence of an A-MSDU is indicated by using the reserved bit (bit 7) in the QoS Control field of the data frame format. All the aggregated MSDUs are of the same TID. Aggregation is allowed for MSDUs whose destination address (DA) and source address (SA) parameter values map to the same RA and TA values. This requirement implies that all the aggregated MSDUs for

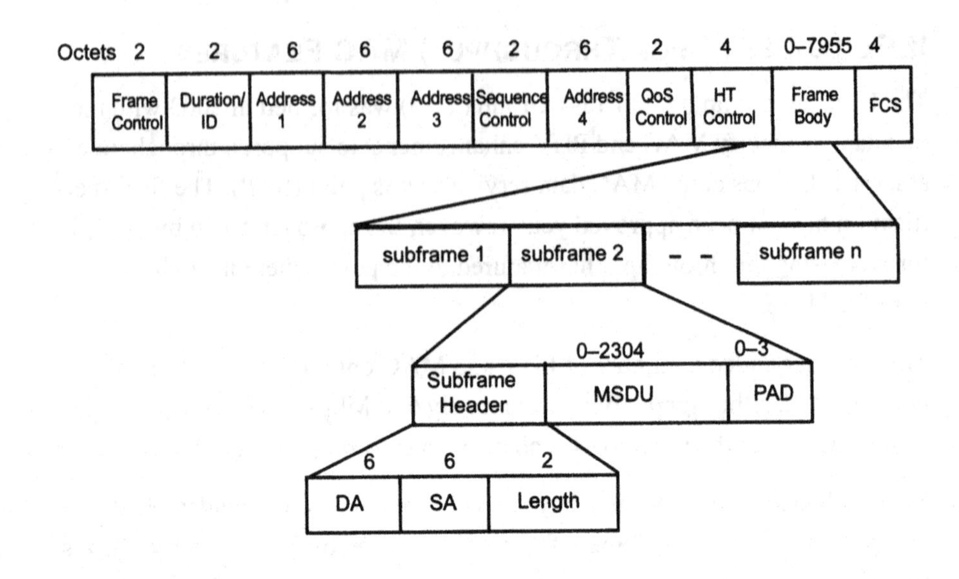

Figure 4–35: A-MSDU Structure

the same receiver are transmitted by the same transmitter. It is also possible to have different DA and SA in subframe headers of the same A-MSDU as long as they all map to the same Address 1 and Address 2 values. The maximum A-MSDU length is set at 3829 bytes or 7935 bytes, as indicated by the Maximum A-MSDU Length subfield of the HT Control.

The A-MPDU format is shown in Figure 4–36. Each A-MPDU subframe is preceded by a 4-byte MPDU Delimiter. The purpose of the MPDU Delimiter is to locate MPDUs within an A-MPDU. The eight-bit CRC (cyclic redundancy check) is used to protect the Reserved and the MPDU Length fields of the MPDU Delimiter. The Delimiter Signature is a pattern that may be used to detect the MPDU Frame Delimiter when scanning for a delimiter.

The maximum length of a MPDU is 4095 bytes. All MPDU subframes within an A-MPDU are destined to the same receiver and have the same Ack Policy as defined by the Ack Policy field in the QoS Control. All MPDU subframes will have the Duration/ID value.

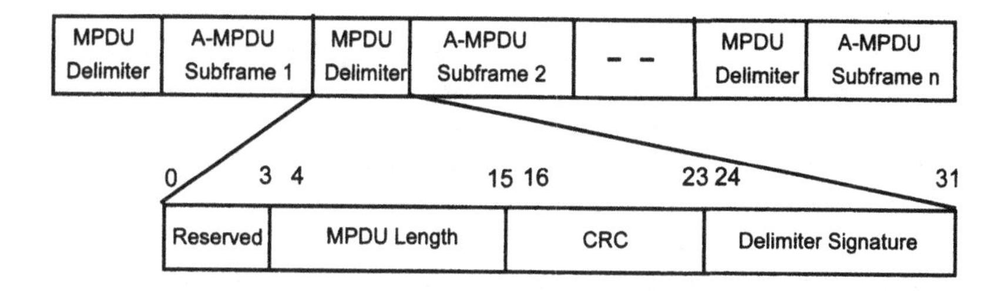

Figure 4–36: A-MPDU Structure

Each station specifies the maximum A-MPDU length that it can receive. When it receives an A-MPDU that is longer than the limit specified, it will accept A-MPDU up to the limit specified and discard the rest of the A-MPDU.

Figure 4–37 shows the throughput achieved with IEEE 802.11n for two cases. The first case assumes 20 MHz with 64-QAM and a coding rate of 3/4. The total rate in this case is 58.5 Mbps. The second case assumes 40 MHz with 16-QAM and a coding rate of 3/4. The total rate in this case is 80 Mbps.

Figure 4–37 demonstrates clearly the crucial role longer MSDU units play in improving the efficiency of the transmission facility. The longer MSDU units require longer transmission time, which in turn reduces the time spent carrying channel and MAC overhead.

IEEE 802.11n Block Acknowledgment

IEEE 802.11n has introduced a number of variations to the basic IEEE 802.11e block acknowledgment. Support of these variations required the changes of the syntax and the semantics of the Ack Policy subfield in the QoS Control field and the BlockAck frame.

IEEE 802.11n introduced the concept of Implicit Ack where a BlockAck is generated by a station without the need to receive a BlockAck Request frame.

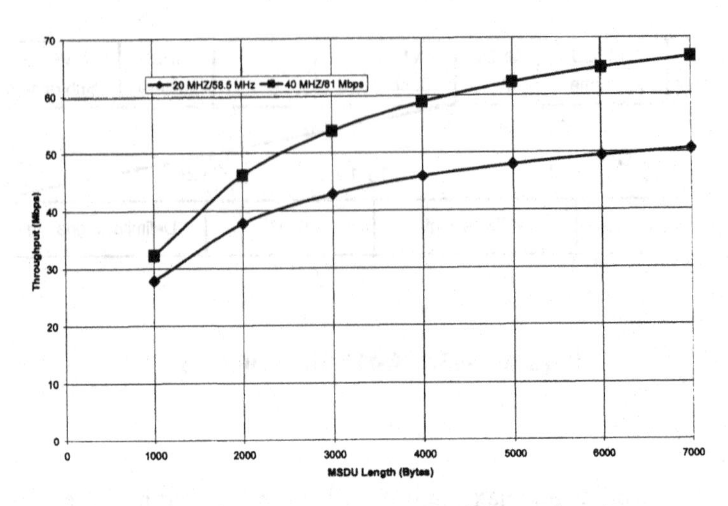

Figure 4–37: IEEE 802.11n Throughput

The support of Implicit Ack is indicated in the Ack Policy of the QoS Control field.

The BlockAck frame has the same format introduced by IEEE 802.11e, as shown in Figure 4–28 on page 132. However, fields of the BA Control field have changes to reflect the new variations introduced by the IEEE 802.11n. The format for the BA Control field is shown in Figure 4–38.

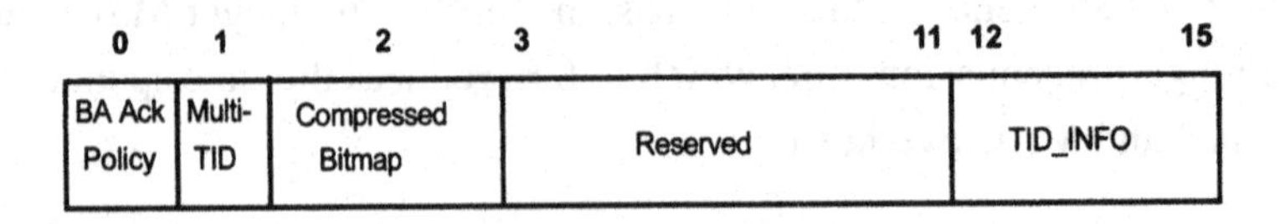

Figure 4–38: IEEE 802.11n BA Control Field

The Multi-TID and Compressed Bitmap bits are used to indicate the variations of the BA supported. These variations are: Basic BlockAck, Compressed BlockAck, and Multi-TID BlockAck. The Basic BlockAck variant uses the same BlockAck format as defined in IEEE 802.11e. The compressed BlockAck variant uses an 8-byte BlockAck Bitmap field instead of the

	MAC Header	IP Header	UDP Header	RTP Header	Speech Samples
Octets	40	20	8	12	Variable

Figure 4–39: VoIP Packet Encapsulation

128 bytes used in the Basic BlockAck. The Multi-TID BlockAck variant includes one or more instances of the per TID information. The number of BlockAck instances is indicated in the TID_INFO field of the BA Control field.

IEEE 802.11n introduces new procedures such as the reverse direction (RD) protocol and power-saving multi-poll (PSMP), which are beyond the scope of this book.

A Simple Engineering Example

The main reason for initiating the work on WLAN QoS is the need to support real-time applications. Among those applications, VoIP is of great interest. This section introduces a simple engineering example in an attempt to predict the voice-carrying capacity of certain IEEE 802.11 technology. The voice carrying capacity is defined as the number of voice calls that can be supported with the transmission rate and the associated overhead.

The VoIP process acquires speech samples from the decoder and encapsulates them using IP protocol suite. Figure 4–39 shows a common encapsulation method using RTP (Real-Time Transport Protocol) and UDP (User Datagram Protocol) for transporting voice frames over IP transport networks.

The amount of speech samples accumulated depends on the voice decoder used and the packetization delay. Assuming G.721 voice coding that generates 64 Kbps speech and 20 ms packetization delay, speech samples occupy 160 bytes. The total size of a VoIP packet is 240 bytes including protocol overhead. A one-direction single voice call will have a rate of 96 Kbps, assuming a single VoIP packet is generated every 20 ms.

Figure 3–25 on page 89 shows that for an MSDU size of 240 bytes and IEEE 802.11a at 54 Mbps, the WLAN throughput is approximately 12 Mbps. The voice carrying capacity can be calculated by dividing the WLAN throughput by two times the rate of a single VoIP connection. The voice-carrying capacity in this case is about 60 calls.

The previous computation assumes 100% loading of the WLAN medium. In practice, a 100% loading is not recommended because of its impact on voice delay and delay variation performance. The exact loading conditions must be determined, taking into consideration delay and loss performance of voice packets. While simulation is needed to fine tune the model and find a suitable loading condition, the simple model presented here is a simple one to compute a first-order approximation for the number of voice calls that can be supported, given a certain IEEE 802.11 technology.

Chapter 5 Interworking with WLAN QoS

Network technologies are rarely deployed in isolation. An information unit, whether a packet or a frame, traverses a number of network technologies from its creation at the source until it is received by the intended destination. Interworking between the different network technologies is needed to ensure consistent application of the same treatment, e.g., priority treatment, as the information unit moves from one network segment to the next.

Luckily a significant amount of consolidation has been done in the telecommunication industry. Two dominant network technologies have emerged. Those based on networks on the Internet protocol are commonly referred to as *IP networks*. Those based on the Ethernet technology are commonly referred to as *Ethernet networks*. Because of the dominance of these two technologies at the enterprise and provider networks, the discussion here will focus on the interworking of WLAN QoS with IP and Ethernet networks.

Interworking between different networking technologies can be performed on different levels. Two common ways for achieving the desired interworking are service interworking and network interworking.

Service interworking is shown in Figure 5–1. Two customer equipment (CE) devices with different interface types are connected to two heterogeneous networking technologies, network type A and network type B, as in Figure 5–1. One CE could be, for example, an IP phone. A session can be set up between the two devices from two different networking technologies. Service interworking involves the translation between the two frame formats supported by the two networking technologies at the two sides of the interworking unit (IWU). The IWU itself might be a logical entity and can be implemented as an add-on function at one of the two networks.

Network interworking is shown in Figure 5–2. In this case the two CEs are of the same type and are connected to networking technologies of the same type, e.g., type A. A different networking technology, type B, is used to connect the

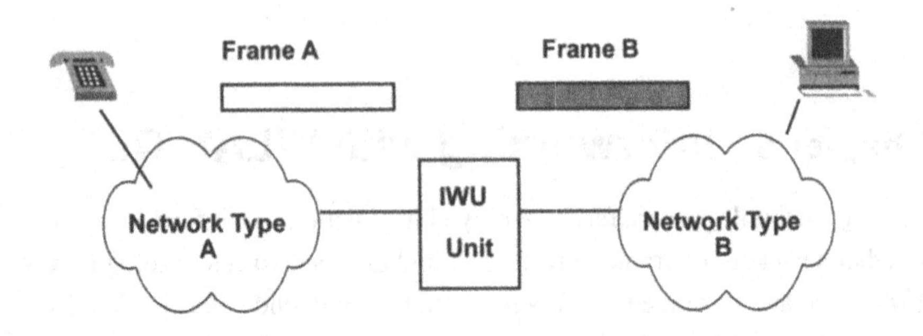

Figure 5–1: Service Interworking

two similar networks, as shown in Figure 5–2. Interworking in this case involves the encapsulation of the frame format of type A inside the frame format of type B. Type A frame is recovered again at the other side of the network by stripping out the encapsulation header.

Interworking between the different networking technologies involves many issues, including routing and signaling. QoS interworking ensures that the QoS of a session does not deteriorate as information units are passed from one networking technology to another.

Whether frame encapsulation or frame translation is used, the QoS interworking depends the mapping of QoS features between two different technologies with possibly different QoS models. It usually involves setting appropriate fields in the frame header as well as setting the right signaling parameters, if necessary.

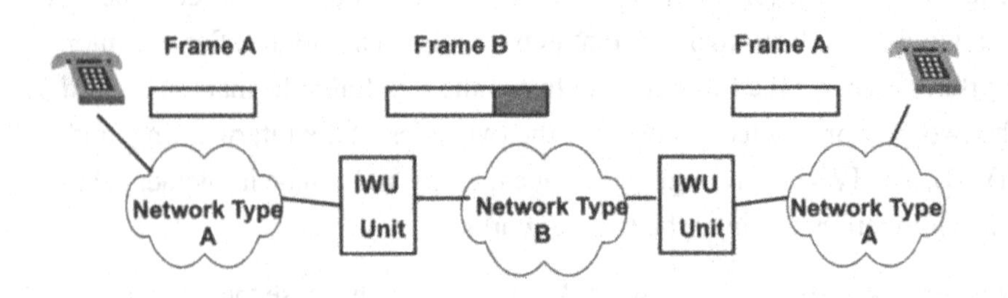

Figure 5–2: Network Interworking

COMMON WLAN DEPLOYMENT SCENARIO

WLAN devices are prevalent in industry, where they are deployed to provide the work force with the ability to move around the enterprise networks without losing connectivity. WLAN devices are also deployed during conventions to provide participants the means to stay connected to their offices. In both cases WLAN is deployed in infrastructure mode where WLAN APs are connected using wired Ethernet switches to provide connectivity to the Internet and the outside world. Figure 5–3 shows the particulars of a common WLAN deployment scenario.

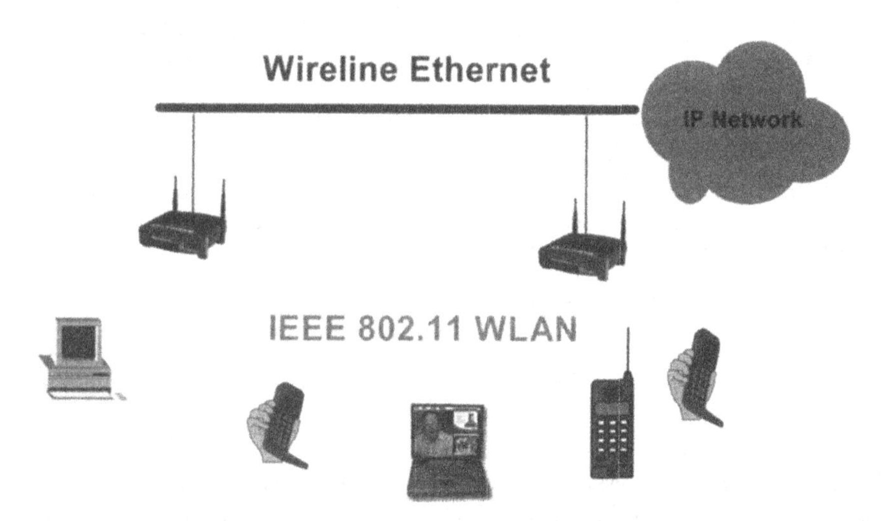

Figure 5–3: Common Interworking Scenario

Depending on the deployment scenario, WLAN devices might be equipped with QoS capabilities as described in Chapter 4. WLAN device capability might support EDCA or HCCA or both. A WLAN frame is generated at the enterprise level and is sent using either an EDCA or HCCA access mechanism. The frame is expected to receive the same kind of treatment as it passes to the wired Ethernet and eventually to the Internet.

WLAN QoS Interworking with Ethernet

Ethernet QoS capabilities are described in Chapter 2. Ethernet supports a differentiating QoS model and relies on user priority (UP) bits to provide different treatment of the different application types. Table 5–1 shows the mapping between the different traffic types and the UP values.

Table 5–1: Priority to Traffic Types Mapping

Priority	Acronym	Traffic Type
1	BK	Background
0 (Default)	BE	Best Effort
2	EE	Excellent Effort
3	CA	Critical Application
4	VI	Video < 100 ms latency and jitter
5	VO	Voice < 10 ms latency and jitter
6	IC	Internetwork Control
7	NC	Network Control

Figure 5–4a shows a possible interworking scenario where WLAN in infrastructure mode provides multimedia service to groups of users that are interconnected using wired Ethernet. Wired Ethernet in this case is used to extend the range of the WLAN segments. Depending on the medium access mechanism employed at the WLAN, two different cases need to be considered.

The first case is when EDCA is used in WLAN for medium access. Figure 5–4b shows the IEEE 802.11 data frame in which the value of the TID subfield of the QoS Control field reflects the UP of the transmitted frame. The value of the TID determines the access category used for this frame as defined in Table 4–1 on page 100. For instance, if the WLAN frame supports voice application, the value of the TID field will be 0110, and AC_VO will be used for channel access. The use of AC_VO will ensure that the voice frame receives preferential treatment and is transmitted with the minimum possible delay.

The QoS interworking in this case between WLAN and wired Ethernet is limited to mapping the value of the TID field to the frame priority bits in the tag control field of the Ethernet frame, as shown in Figure 5–4c. No further action is needed. Interworking is simplified by the fact that WLAN employing EDCA has the same QoS model as the wired Ethernet.

The second case that needs to be considered is when WLAN is employing HCCA for channel access. QoS interworking becomes more challenging in this case. The challenge stems from the fact that WLAN employing HCCA and wired Ethernet have different QoS models—a reservation model at the WLAN side and a differentiating model at the wired Ethernet.

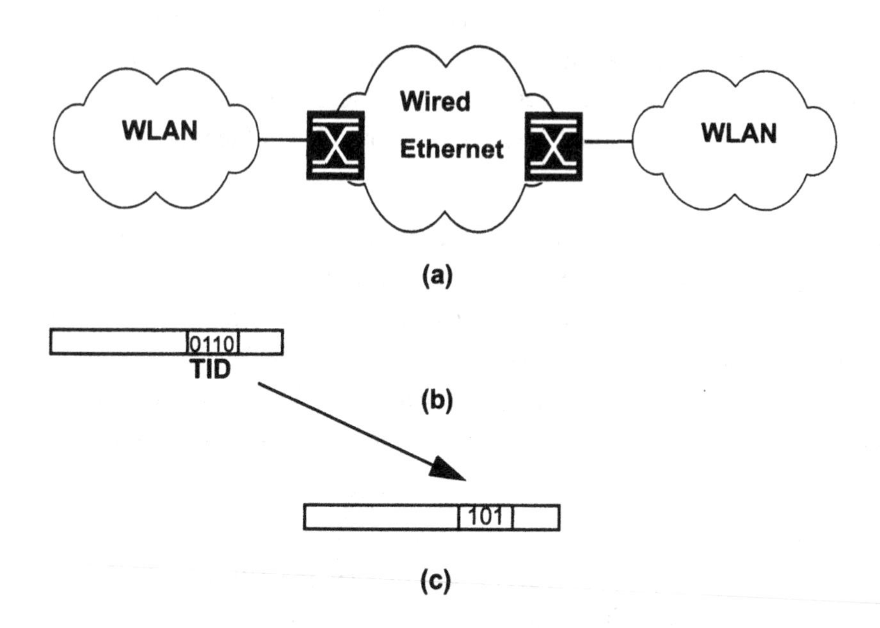

Figure 5–4: WLAN and Wired Ethernet Interworking

What adds to the difficulty of the problem is that no direct and clear way exists to set the priority values at the wired Ethernet, based on information contained in the IEEE 802.11 frame. With HCCA as the channel access mechanism, the TID subfield of the QoS Control field no longer reflects a UP value but is rather used to indicate the TSID. TSID has no significance at the wired Ethernet.

Even though QoS interworking in this case is challenging, there are ways to support the interworking. They usually require intercession from the network administrator. The network administrator in this case can install configuration values or policies to be used to achieve the desired interworking. Installing proper configuration values is straightforward when the different network segments are managed by the same entities. Otherwise, the different administrations need to agree on the configuration values.

One method to deal with the interworking in this case relies on the assumption that HCCA reservation per traffic stream and the associated contention-free polling of stations are introduced mainly to satisfy application delay requirements. It can then probably be assumed that HCCA traffic streams are supporting real-time applications such as voice and video. Therefore, the interworking in this case reduces to mapping HCCA streams as identified by the value of the TID subfield to either voice or video traffic type, as shown in Table 5–1 on page 148.

If HCCA is used as the only access mechanism for all traffic types, QoS interworking can still be possible if the network administrator defines certain TSID to be used for each traffic type mentioned in Table 5–1. For instance, a data session at WLAN is required to use 1101 as its TSID. The QoS interworking then reduces the mapping between the TSID values and the UP values in the IEEE 802.1 frame header in the same way as shown in Figure 5–4.

It is worth noting that with HCCA as the access mechanism, resources are reserved per traffic stream at the WLAN segment. Resource reservation is not possible at the wired Ethernet segment. However, the network administrator might configure a certain amount of resources for each traffic type. The desired configuration can be achieved by implementing a scheduling algorithm at the wired Ethernet segment that can be controlled by setting different scheduling weights for different traffic types. The network administrator can then set admission limits for each traffic type at the WLAN segment so that the configured resources at the wired Ethernet are sufficient to handle the expected aggregate load.

This description outlines the interworking functions needed in the direction from the WLAN to the wireline Ethernet. Interworking in the other direction from the wired Ethernet to WLAN is achieved by mapping from the Ethernet UP in the Ethernet frame header to WLAN TID subfield in the QoS Control field following the same rules.

Important to note is that the wired Ethernet need not know whether EDCA or HCCA media access is employed at the WLAN side.

WLAN QoS Interworking with IP Networks

The IP QoS capabilities are described in Chapter 2. IP Networks could support either a differentiating model employing the IETF Diffserv architecture or a reservation model employing the IETF IntServ architecture and its related service types.

Figure 5–5 shows a WLAN-IP interworking scenario where two WLAN sides are connected together by IP routers. Routers can have the WLAN AP implemented on one of their interfaces, or the WLAN AP can be connected to IP access routers. Unlike interworking with wired Ethernet, more than two cases must be considered. Independent of the media access mechanism used at the WLAN side, there is the need to consider the two possible IP QoS models.

Figure 5–5: WLAN-IP Interworking Scenario

The first case to consider is when EDCA is used at the WLAN side and Diffserv is used at the IP side. The QoS models used at the two sides are compatible, and interworking should be straightforward. This expectation is largely true. However, care must be taken at the IP side because the IP Diffserv architecture has no concept of service or traffic types as those expressed in Table 5–1 and applicable for WLAN and wired Ethernet.

Because defining those Diffserv services is not within the scope of this book, we make use of the general guidelines that are presented in Babiarz et al. [B3]. Table 5–2 is reproduced from Babiarz et al. [B3]. It shows how the Diffserv PHB may be used to construct service types that support different applications with different requirements.

Table 5–2: DSCP to Service Class Mapping

Service Class Name	DSCP Name	Application Example
Network Control	CS 6	Network routing
Telephony	EF	IP telephony bearer
Signaling	CS 5	IP telephony signaling
Multimedia Conferences	AF41, AF42, AF43	H.323/v2 video conferences (adaptive)
Real-Time Interactive	CS 4	Video conferences and interactive gaming
Multimedia Streaming	AF31, AF32, AF33	Streaming video and audio on demand
Broadcast Video	CS 3	Broadcast TV and live events
Low Latency Data	AF21, AF22, AF23	Client/server transactions web-based ordering
OAM	CS 2	OAM&P
High Throughput Data	AF11, AF12, AF13	Store and forward applications
Standards	DF (CS 0)	Undifferentiated applications
Low Priority Data	CS 1	Any flow with no BA assurance

Based on Table 5–2 and the WLAN traffic types, Table 5–3 shows a possible mapping between the UP bits at the WLAN side and the DSCP at the IP side. However, the table shows one mapping example. It does not imply in any way that this is the only mapping.

Table 5–3: Ethernet Priority to DSCP Mapping

Priority	Acronym	DSCP
1	BK	CS 1
0 (Default)	BE	CS 0
2	EE	AF11, AF12, AF13
3	CA	AF21, AF22, AF23
4	VI	CS 3
5	VO	EF
6	IC	CS 5
7	NC	CS 6

The second interworking scenario to be considered is where EDCA is used at WLAN, and IntServ is used as the QoS model implemented at the IP. IntServ supports two service types, guaranteed service (GS) and controlled load service (CLS), in addition to the normal IP best-effort (BE) service as described in Chapter 2 in "Guaranteed Service (GS)" on page 49. WLAN traffic types have to be mapped to one of these three service types. For example, voice traffic type could be mapped to the GS service, excellent effort (EE) can be mapped to CLS, and Ethernet best effort to IP BE service.

RSVP signaling protocol is used at the IP network for session establishment. Because EDCA does not include the concept of a traffic stream, it is possible to perform the mapping described here between WLAN EDCA and IP IntServ on an aggregate basis. With aggregation, only one GS session needs to be established. Voice traffic at the WLAN side will be aggregated and assigned to the established GS session, as shown in Figure 5–6.

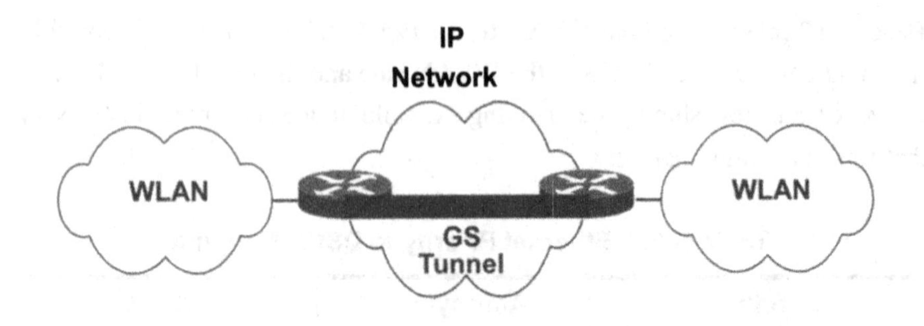

Figure 5–6: Service Aggregation at the IP Network

Session establishment at the IP side will require the specification of the RSVP TSPEC parameters. Those parameters are *p*, *r*, *L*, *m*, and *M* as described in Figure 2–2 on page 48. These parameters can directly be extracted from the WLAN parameters, as shown in Table 5–4. The sigma, or sum sign (Σ), in Table 5–4 reflects the fact that the session established at the IP network supports the aggregate sum of certain traffic types, as explained previously.

Table 5–4: RSVP TSPEC Parameters

RSVP TSPEC Parameters	WLAN Aggregate Parameters
p	Σ Peak Data Rate
r	Σ Nominal Data Rate
b	Σ Maximum Burst Size
m	Configured
M	Maximum WLAN Frame Size

The next interworking scenario to be considered is when HCCA is used for medium access at the WLAN. The Diffserv QoS model is used at the IP network. This scenario is similar to the interworking with wireline Ethernet described in the previous section. The same rules are apply here, where the network administrator relies on predefined configurations of TSID values and Diffserv DSCP mapping.

The last scenario to be considered is when HCCA is used at the WLAN and IntServ is used at the IP network. In this case the QoS models at both sides of the network are compatible in the sense that they both are classified as reservation models. A traffic stream established using HCCA at the WLAN requires a particular reservation level based on the traffic parameters included in the TSPEC IE. Those parameters are extended through the IP network by using RSVP session establishment procedure to set up a session with the appropriate reservation and appropriate service type, e.g., GS or CLS. The mapping between the WLAN TSPEC parameters and the RSVP Source Tspec parameters follows the same rules as in Table 5–4. However, in this case the sum sign (Σ) is not needed because one RSVP session is established for each WLAN TSID.

Even though an aggregate session is not a requirement in this case, a network administrator might still choose to establish aggregate sessions at the IP network for operational or scalability reasons.

Chapter 6 Concluding Remarks

An objective of this book is to provide access to the WLAN QoS MAC enhancements in an simple and clear way. To achieve this objective, the book introduces the subject of QoS and traffic management and the associated definitions, mechanisms, architectures, and examples. The QoS subject has been presented in this way to position the WLAN QoS enhancements in the right perspective and to show the rationales of particular selections. In that respect, this book can be used to gain a quick overview of the QoS subject and the related issues.

The subject of QoS and traffic management usually requires a certain level of mathematical background. The description presented in this book avoids the need for rigorous mathematics for the sake of clear presentation. It is hoped that this approach will be adequate for most readers. Those readers who are willing to examine the subject more carefully can refer to the references

Planning network resources to achieve the desirable levels of performance requires careful engineering of the network resources, such as buffers and transmission facilities. For networks supporting multimedia applications, design and engineering of those traffic management mechanisms—such as transmission schedulers and buffer-sharing algorithms—become very important. They ensure that each multimedia class acquires its allocated resources and achieves its performance objectives. The book discusses simple engineering examples related to admission control and capacity planning that should prove useful to the reader.

Interworking between networks at the QoS level is a subject that have not received adequate attention in the published literature. This book is unique in having a separate chapter dedicated to this subject. The scope of the book limited discussion to the issues related to interworking between WLAN QoS, IEEE 802.1 QoS, and IP QoS features. The subject of interworking extends far beyond the limited discussion presented in this book. It is hoped that the discussion presented here will raise the awareness of this subject's importance and will help network planners and engineers support end-to-end services that span multiple network technologies.

In discussing the subject of interworking, it was necessary to provide background related to the QoS features and mechanisms proposed and implemented for IEEE 802.1 Ethernet bridging and IP technologies. This book can serve as the starting point towards understanding these topics.

Finally, experience has shown that deploying QoS features for any technology is usually a slow process that network providers approach with caution. One of the objectives for writing this book is to hasten the deployment of the QoS features in the WLAN. It is hoped that this book will be a small step towards the achieving this goal.

Bibliography

[B1] AF-PNNI-0055.002 (2002) Private Network-Network Interface Specification Version 1.1, ATM Forum Technical SPecification, April 2002.

[B2] AF-TM-0121.000 (1999) Traffic Management Specifications: Version 4.1, ATM Forum Technical Specification, March 1999.

[B3] Babiarz, J; Chan, K; and Baker, F. Configuration Guidelines for DiffServ Service Classes, IETF RFC 4594, September 2006.

[B4] Black, S. An Architecture for Differentiated Services. IETF RFC 2475, December 1999.

[B5] Braden, R. Editor. Resource ReSerVation Protocol (RSVP) Version 1 Functional Specification, IETF RFC 2205, September 1997.

[B6] CCITT I.370. Congestion Management for the ISDN Frame Relaying Bearer Service, CCITT Recommendation, 1991.

[B7] Davie, B. Editor. An Expedited Forwarding PHB (Per-Hop Behavior), IETF RFC 3246, March 2002.

[B8] Demichelis, C. and Chimento P., IP Packet Delay Variation Metric for IP Performance Metrics (IPPM), IETF RFC 3393, November 2002.

[B9] Floyd, S. and Jacobson, V. Random Early Detections Gateways for Congestion Avoidance, IEEE/ACM Transactions on Networking, V.1 N.4, August 1993.

[B10] Gerla, M and Kleinrock, L. Flow Control: A Comparative Survey, IEEE Transactions on Communications, Vol. COM-28, No. 4, pp. 553-574, April 1980.

[B11] Gibbens R. and Hunt P. Effective Bandwidth for the Multi-Type UAS Channel, Queueing Systems, Vol. 9, 1993.

[B12] Goyal, P; Lam, S; and Vin, H. Deterministic End-to-End Delay Bounds In Heterogeneous Networks, Proceedings of the 5th International Workshop on Operating Systems Support for Digital Audio and Video, 1995.

[B13] Heinanen, J; Baker, F; Weiss, W, and Wroclawski, J. Assured Forwarding PHB Group, IETF RFC 2597, June 1999.

[B14] IEEE Standard 802.1ad™ (2005) Virtual Bridged Local Area Networks—Amendment 4: Provider Bridge, May 2005.

[B15] IEEE Standard 802.1Q™ REV (2005) Virtual Bridged Local Area Networks- Revision, May 2005.

[B16] IEEE Standard 802.11™ (1999) WIreless LAN Medium Access Control (MAC) and Physical Layer (PHY) Specification, 1999.

[B17] IEEE Standard 802.11e™ (2005) WIreless LAN Medium Access Control (MAC) and Physical Layer (PHY) Specification- Amendment: Medium Access Control (MAC) Quality of Service Enhancements, 2005.

[B18] Jacobson, V. Congestion Avoidance and Control, Proc ACM SIGCOMM 88, August 1988.

[B19] Jacobson, V. Towards Differentiated Services for the Internet. Presentation to Bay Networks Architecture Lab, September 1997. Available at: ftp://ftp.ee.lbl.gov/talks/vj-baydsarch.pdf.

[B20] Jun, J; Peddabachagari, P; and Sichitiu, M. Theoretical Maximum Throughput of IEEE 802.11 and its Applications, Proceedings, 2nd IEEE International Symposium of Networking Computing and Applications, 2003.

[B21] Kleinrock, L. Queueing Systems: Vol. 1: Theory, John Wiley and Sons, 1975.

[B22] Kleinrock, L. Queueing Systems: Volume II: Computer Applications, John Wiley, 1975.

[B23] Kuehn, P. Multiqueue Systems With Nonexhaustive Service, Bell Systems Technical Journal, Vol. 58, March 1979.

[B24] Le Boudec, J. and Thiran, P. Network Calculus: A Theory of Deterministic Queueing Systems for the Internet. On-Line Version of Springer Verlag LNC 2050. May 2004. Available at:
http://ica1www.epfl.ch/PS_files/netCalBookv4.pdf.

[B25] MEF 10.1 (2006) Ethernet Service Attributes Phase 2. Technical Metro Ethernet Forum Technical Specification, November 2006.

[B26] Nichols, K; Blake, S; Baker, F; and Black, D. Definition of the Differentiated Services Field (DS Field) in the IPv4 and IPv6 Headers, IETF RFC 2474, December 1999.

[B27] Padhye, J, Firoiu, V., Towsley, D, and Kurose, J, Modeling TCP Reno Performance: A Simple Model and its Empirical Validation, IEEE/ACM Transactions on Networking, Vol. 8, April 2000.

[B28] Parekh A and Gallager R. A Generalized Processor Sharing Approach to Flow Control in Integrated Services Networks: The Single-Node Case, IEEE/ACM Transactions on Networking, Vol. 1, June 1993.

[B29] Ramakrishnan, K; Floyd, S.; and Black D. The Addition of Explicit Congestion Notification (ECN) to IP, IETF RFC 3168, September 2001.

[B30] Shenker, S; Partridge, G; and Guerin, R. Specification of Guaranteed Quality of Service, IETF RFC 2212, September 1997.

[B31] White, P. RSVP and Integrated Services in the Internet: A Tutorial, IEEE Communications Magazine, May 1997.

[B32] WiFi Alliance WMM Specification Version 1.1.

[B33] Wroclawski, J. Specification of the Controlled-Load Network Element Service, IETF RFC 2211, September 1997.

[B34] Wroclawski, J. The Use of RSVP with IETF Integrated Services, IETF RFC 2210, September 1997.

Index

V

W

Printed in the United States
By Bookmasters